HYPE

HYPE

A DOCTOR'S GUIDE TO MEDICAL MYTHS,

EXAGGERATED CLAIMS

AND BAD ADVICE—HOW TO TELL

WHAT'S REAL AND WHAT'S NOT

NINA SHAPIRO, M.D.

with KRISTIN LOBERG

St. Martin's Press ☒ New York

www.stmartins.com

Designed by Steven Seighman

Library of Congress Cataloging-in-Publication Data

Names: Shapiro, Nina, 1947– author. | Loberg, Kristin, author.
Title: Hype : a doctor's guide to medical myths, exaggerated claims and bad
    advice — how to tell what's real and what's not / Nina Shapiro, M.D., with
    Kristin Loberg.
Description: New York : St. Martin's Press, [2018] | Includes bibliographical
    references and index.
Identifiers: LCCN 2017055072| ISBN 9781250149305 (hardcover) |
    ISBN 9781250149312 (ebook)
Subjects: LCSH: Medical misconceptions. | Medicine, Popular.
Classification: LCC R729.9 .S55 2018 | DDC 610—dc23
LC record available at https://lccn.loc.gov/2017055072

Our books may be purchased in bulk for promotional, educational, or business use. Please contact your local bookseller or the Macmillan Corporate and Premium Sales Department at 1-800-221-7945, extension 5442, or by email at MacmillanSpecialMarkets@macmillan.com.

First Edition: May 2018

10 9 8 7 6 5 4 3 2 1

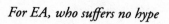
For EA, who suffers no hype

CONTENTS

WORRYWARTS AND FEARMONGERS

The Dangers of Magical or Misinformed Thinking

DOES SUGAR FEED CANCER CELLS?

WHEN CAN NATUROPATHIC MEDICINE HAVE MERIT?

SHOULD YOU BE WORRIED ABOUT GENETICALLY MODIFIED FOODS?

CAN VACCINES TRIGGER BRAIN DISORDERS SUCH AS AUTISM?

At least a couple of times a week, I encounter misinformed patients whose ideas about what's "good" for them—or a family member—are totally wrong. Or they ask questions about what they've heard about in the media or read online and are surprised by my answers, which often contradict conventional public wisdom. Some examples: *My child stopped eating dairy and he doesn't get sick anymore. . . . I only get the flu when I get a flu shot. . . . Did you know that throat lozenges can cure strep throat? I read that in the newspaper.*

In my work at UCLA as a surgeon, and as someone who has had her feet firmly planted in both clinical and academic medicine for more than twenty years, I help patients and their families make decisions every day about their health. I find myself dispelling a lot of myths. We live in times filled with suspicion: Every day the media delivers swarms of health-related information that can swiftly trigger fear or inspire us to change our habits overnight. From the headlines about brain-eating amoebas in freshwater lakes to the alleged ills of gluten, sugar, vaccines, foods from genetically modified organisms (GMO), and the chance of getting cancer from tap water, the onslaught of news can be downright overwhelming—and is potentially *harmful*. One day coffee is good for you and protective against dementia, the next day it's declared a potential carcinogen.

Claims that routinely circulate are frequently overblown, misleading, based on shoddy research, or completely false. Some of the ideas are hocus-pocus, created to prey on the vulnerable. Common offenses include exaggerating the benefits of many vitamins, herbs, supplements, homeopathy, skin creams, antiaging schemes, cold remedies, and unconventional anticancer programs. Talented scammers know how to lead the public into believing in causation when there is none, building hype for everything from arnica to zinc.

Unfortunately, many individuals don't know where to turn for unbiased, trustworthy advice, and the ease with which misinformation is propagated on the internet leaves people's heads spinning. Hence the purpose of this book.

I started this book several years ago. After the 2011 publication of *Take a Deep Breath,* which spoke to parents who have children with breathing problems, I realized I needed to write a much broader book that speaks to people about their health decisions on a wide array of issues. I would argue that more hype and misinformation exists in medicine and health circles than in any other area. I say this not just as a physician, but also as a mom, wife, sibling, daughter, and someone's friend. I am fortunate to live in a world surrounded by so many smart people who are successful in a multitude of ways. Some are talented actors, directors, or producers; some are innovative entrepreneurs; many are far better organized and far more engaged parents than I could ever hope to be. But what I notice so often is a huge disconnect. *Hypocrisy* might be too strong a word, but I wouldn't rule it out.

When it comes to health, well-being, nutrition, and medical choices, well-educated people can have some pretty funny ways of thinking. No doubt personalities and emotions come into play, especially when such decisions are being made for one's own body or for a spouse, parent, or child. Health choices can be downright bizarre. At my kids' school, for instance, I would frequently hear parents concerned, to the point of obsession, about hand washing. Yes, this is important, don't get me wrong. Yet these same parents don't immunize their children and rally to ban any Styrofoam or plastic while driving a gas-guzzling SUV. Some people worry about eating inorganic foods with dyes and artificial colorings but will text and drive and skip the seat belt on occasion.

Surgeons have a unique way of thinking—different, even, from that of doctors in other specialties. Every decision we make involves risks and benefits, and our decisions often have to be made in an instant. We think and do almost simultaneously. As the saying goes, a good surgeon knows how to operate; a great surgeon knows when not to. One choice is not necessarily more aggressive than the other. We see how bodies work in live, real time. We see the internal damage that no physical exam, blood test, or highest-level X-ray can replicate.

We think and act fast. Many view surgeons as erratic and impulsive. In many ways, this is true. If the sh*t is hitting the fan, our impulses kick in, and we act on impulse to save a patient, stop the bleeding, open an airway, clamp a vein. We're not always polite when a patient is, as we call it, circling the drain, crashing, or trying to die. Accept our apology.

We're not just butchers. We know how the human body works, metabolizes, absorbs, digests, just as much as nonsurgeons do. We know that rubbing cream on your skin will rarely have a direct effect on other organs. We know that drinking a juice marketed to boost your immune system will not do that. We know that the liver is a great detox center, the kidneys are a terrific filter, and the brain is the most powerful organ. We are so keenly tuned in to the ins and outs of the body that when a patient of a certain age, weight, and medical status has a specific operation for a specific length of time and will not be eating or drinking for a certain period afterward, we can calculate to the drop of water—milliequivalent of each electrolyte—and milligram of protein how much that patient's body will need via intravenous fluids to keep the patient in perfect fluid and electrolyte balance indefinitely. Undergoing abdominal surgery is more taxing on the body than running a marathon, climbing a mountain, or sweating through a day of hot yoga, yet we can "get it right" when it comes to ideal fluid and electrolyte balance.

Sports drinks, protein boosts, and energy bars are all guesswork. They go through the gastrointestinal tract and may not be absorbed in the right amount by the right organ. Just as postsurgical patients can't by drinking liquids maintain their fluid/electrolyte balance, people ingesting supplements, drinks, powders, and bars oftentimes get no benefits because of where those substances end up. (Hint: Check your pee and poop.) When one receives intravenous fluids, the contents directly enter the vascular system, going from the veins through the heart, to the arteries, and then to the organs of the body, where absorption of fluid and electrolytes takes place. Some vessels go to the kidneys, which filter out the unnecessary fluid and electrolytes, so we can pee them out. When you eat something, the contents go from the mouth to the esophagus to the stomach, and then the intestines. After this, some of the contents get pooped out. Which of these make it past the intestines, the liver, and the kidneys is much more guesswork than in the absorption of intravenous fluid contents.

I admit that we surgeons are arrogant, but we know our limits, hopefully before we enter the operating room. We know that we don't know everything, but we also know that the latest and greatest new procedure, technology, or technique may be based on great advertising but not great science.

I come from a family of doctors, so I had an early exposure to the medical profession.

When I was a freshman in college, I volunteered in an emergency room during winter break. This was back in the days before privacy restrictions and age minimums in hospitals. I got to put in some stitches, drain some pus, and learn about hardships that so many suffer on a daily basis, especially the homeless and the drug dependent. My dad distinctly remembers coming to pick me up one day but he couldn't find me. Until he saw I was gloved, masked, and gowned in the middle of a gunshot-wound case—looking calm, no less. It was as chaotic as a scene from a TV drama. But to me it was unremarkable. After that day, my dad knew there'd be another surgeon in the house.

I went to Harvard Medical School, where, even in the late twentieth century, few women braved surgical fields. I loved surgery, and my specialty—otolaryngology, which focuses on the ears, nose, and throat—offered the perfect mix of healthy and sick patients, young and old, surgical and outpatient work. It was like no other field in medicine. After all, who else in medicine or surgery really knows the airway? The sphenoid sinus? The temporal bone? The stapes? The cricoid? We got to learn the mysterious stuff lurking in dark corners.

During my early training, I gravitated toward kids and families. It's thrilling to be able to perform the tiniest of surgeries to change a child's life. My specialty is often the "last resort" in an airway emergency, when all other specialties, including anesthesiology, are at a loss. Everything remains routine . . . until it's not. The low-grade sense of imminent terror is always there. There's a lot of waiting, but then for just a few piercing seconds, it's all you.

Medicine has changed tremendously since I first became a doctor more than a quarter of a century ago. In many ways, that change has been for the better. Technology has hastened the rate of development of medical advances. A wider range of new knowledge, not just for patients but for doctors, is available at the touch of a button. Gone are the days of looking up historic or groundbreaking journal articles and textbook chapters in our library's dusty card catalog. Medical students and residents live off their tablets, access PubMed.gov with the utmost facility, and can quote the latest and greatest before we, their crusty old mentors, finish our question. Oftentimes, the article is quoted on CNN before we even get to it. Everything seems easier. This can be good, but not always.

Patients are also different. Again, this is mostly for the good. Access to information is overwhelming. Patients ask more focused, educated, pointed questions and know details about medical care that most doctors wouldn't have known in the past. But this knowledge can often be muddied. There is no "peer review" for the internet, so opinions and falsehoods get lumped together with facts. This is a negative, no matter how you slice it.

My style of practice has evolved over the past twenty or so years. I am now older than most parents. Former patients are becoming parents, and I'm even older than some grandparents. I haven't "seen it all," but I've seen a lot. Most of it has been good: the relief of wellness; the curing of acute or chronic illness. It can be life changing, not just for the patient, but for an entire family—and for the doctor. This is why we stay in medicine. My colleagues and I have had our share of Felixes, as we call them. Like cats, these folks seem to have nine lives. Some are so sick, on death's door so many times, yet pull through and thrive. I've also seen disasters so devastating that I've questioned if I could go on as a doctor. Though rare, in these situations I've watched people die unexpectedly in the OR like a scene from *ER* or *Grey's Anatomy,* as neither I nor any of the experts around me can do anything.

I know I am not alone in this experience. Many of my colleagues take care of the very sick, whether it is from a chronic illness, an acute one, or an accidental trauma. Most of us who share in these experiences need only to mention a first name, or even a calendar date. The common look of understanding means that we need not speak these unspeakable details, but we remember them vividly. This is why many leave medicine.

Working at a place such as UCLA is a bit like being in Oz. We have many wizards, although these folks are real, doing incredible work. We have even more who are seeking something, just like Dorothy and her friends. They may not know, at the outset, what that missing piece might be, but countless physicians, students, and nurses are looking for something. Some may be looking for more technological advances—robotic surgery, robotic laser surgery, remote robotic laser surgery. Some are looking for a new drug therapy with fewer side effects during cancer treatment; prevention of hypoxic brain injury in newborns; smoother pathways in and out of the outpatient surgery unit; and higher success rates in multi-organ transplantation.

What I love about working at UCLA is that no two days are the same. My practice is filled with thousands of interesting people and stories, challenging situations, and new puzzles. You'll be reading about some of them in this book.

HEALTH IS AS HEALTH DOES

I marvel at the changing trends in the definition of "healthy." I remember in the seventies when spaghetti all of a sudden became *pasta*. It somehow seemed healthier, so we ate more of it. When bran muffins (all five hundred calories of them) became health food alongside granola, some people packed on a few more pounds. What followed was the low-fat, low-sugar, sugar-free, no-sugar-added, fat-free, low-calorie craze. The waters are now further clouded by words such as *organic, natural, gluten-free, caffeine-free, nutrient-rich, GMO-free* (although you should know that a little genetic modification can be your friend, as you'll soon find out).

I give out ice pops to kids right after they've had minor surgery. I love to tell them, in front of their parents, that we have orange, purple, and red. They have 100 percent artificial coloring and flavoring and are packed with all kinds of sugars. They contain 0 percent fruit juice. It will be the best ice pop they'll ever have. I often see the parents requesting a second one for their child and downing it themselves. Are these treats any less "healthy" than the ones touting their contents to be packed with real fruit and vitamins, and "no added sugar"? Perhaps they are, but so infinitesimally so that the real benefit of the ice pop (frozen water, for Pete's sake!) is hardly lost when having the cheap stuff. And the cheap stuff probably tastes better, too.

Besides relating to food, the concept of health continues to change in interesting ways. For some reason, we all want to live to 150. Someday, a few of us just might. Genetic testing and potential modification can give us a window into longevity, disease-risk profiles, and long-term prevention. Many of us are trying to get it just right: An aspirin a day or not? Starting when? Hormone replacement therapy or not? Statins to prevent a heart attack? Vitamin supplements, and if so, which ones? B_{12} gives you energy. It also gives you pimples. Vitamin E prevents cancer. It may also cause it. Exercise is good, but which kind is the best? And how much is too much, when we start to pay the price of extremes with joint injuries, chronic pain, and even kidney and heart problems?

I'll be honest: Patients do sometimes get on my nerves with their questions starting with "Do you believe in . . . ?" I wish to interrupt with "The Easter Bunny? Ghosts? The Tooth Fairy?" All of these are, indeed, beliefs. But medicine is not a belief. I practice based on evidence. This is not to say that all medicine is fact. Far from it. Many treatments we use today may, thirty years from now, turn out to have been useless or, worse, harmful. But we give these treatments based on large bodies of evidence that they are helpful. We can, at times, use experimental treatments, but they need to be presented as such. At the least, we need to have a remote understanding of how they work. And, no, magic doesn't count as a reason.

I get asked about alternative medicine and homeopathy all the time. I can't definitively say that these practices are bad or harmful. But when I am asked whether flaxseed oil in the ear canal is good for ear infections, I respond that I've never heard or seen it work, but if you have information from your doctor about its mechanism, I'd love to see it. Or a study. Or something. If a patient was given the homeopathic remedy causticum for a rare bacterial infection, I'd hope that the prescribing caregiver had thought about the mechanism of action. Not that it just sounded caustic. Indeed, it is a dilute form of potassium hydroxide, or slaked lime. Good for lawns. Not so much for atypical bacterial infections.

The purpose of this book is to wake you up to the truth about popular health advice and offer a thoughtful, reliable guide for becoming a smart health consumer and patient. We're going to cover lots of territory, beginning with conspiracy theories that abound in health circles and important concepts to understand such as risk management, causation versus correlation, and what is a good, reproducible, well-designed study. Then we'll move into answering the questions that burn holes in people's heads today: Is there a scientifically proven best diet? Is gluten that bad? Can detoxes be toxic? Does sugar feed cancer cells? When is *organic* code for "outrageous"?

I'll take you on a tour of complementary alternative medicine, showing you when and how it works. I'll also cover the risks related to vaccines, homeopathic supplements, and whether tests such as mammograms, colonoscopies, and DNA screening are overly hyped. I'll get to some of the most popular habits that people keep (or might want to) that can have unintended consequences. I'll answer the following questions: Which vitamins have value? What's the best antiaging secret that's not hype? What's the perfect amount and type of exercise? I aim to make this an enjoyable read, and I trust you will not only find many of my ideas and recommendations useful and surprising, but also that they may relieve you from so much worry in your daily life. Optimal health is easier to achieve than you think. It may always be a moving target, but that doesn't mean you can't enjoy the ride.

All I ask of you at the start is to proceed with an open mind and to cast any preconceived notions aside. We have become a nation obsessed with health issues, so how does one know what is real and what is hype? That's what this book is going to reveal.

HYPE

A SITE TO BEHOLD: THE WILD WEST OF INTERNET MEDICINE

How to Yelp Your Doctor, Check Your Symptoms, and Google If You're Dying

DOES THE INTERNET HAVE QUALITY HEALTH INFORMATION?

HOW DO YOU KNOW IF A WEBSITE IS LEGITIMATE?

ARE ONLINE DOCTOR RATINGS VALUABLE?

WHAT CAN YOU DIAGNOSE ONLINE?

Not too long ago, a friend emailed me late one evening: "Just wondering—my daughter [about fifteen months old] pooped out white poop—I looked it up and it's either something she ate or she has acute liver failure and needs to be rushed in to surgery. What should we do?"

Well, he had already looked it up, which saved me some time, as neither of these possibilities was in my knowledge toolbox. I hadn't thought about other people's poop since my own kids were in diapers, and even on those poops I tried not to focus. While I reassured this pal, via email, since nobody talks to anyone anymore, that this was likely nothing to worry about, I was struck by his well-prepared differential diagnosis of white poop, all from the touch of a button. No, he's not a gastroenterologist, medical resident, or surgeon. He's a movie director.

But in under three minutes, he learned the possible causes of white poop. Another friend, who *is* a doctor, tragically lost her eleven-year-old dog by sudden death. "I'm not sure how this happened. Dogs just don't die like that," her vet declared. But this one did. At the touch of a button, this doctor for humans quickly learned about the potential causes of sudden death in dogs. My nine-year-old son wanted to know the score of a recent Golden State Warriors game. He doesn't have his own computer, so he asked Alexa, Amazon's version of a voice-activated internet knowledge base.

Our internet-based access to knowledge is boundless. From poop to death to basketball, nary a question cannot be answered somewhere online. This has changed the world as we knew it just one to two decades ago, and the world will continue to be changed by the ever-expanding growth of this technology. On a smaller, albeit pretty important, scale, it has altered the way medicine is learned and practiced, the way health care is delivered, and how non-medically-trained individuals access knowledge about their own health and that of their families, including their pets.

When I first became a member of our institution's interdisciplinary craniofacial team, which consists of surgeons, geneticists, nurses, dentists, orthodontists, speech pathologists, audiologists, and social workers, who work together to treat children and families with complex facial abnormalities, we always had a copy of the venerated *Gorlin's Syndromes of the Head and Neck* at the workstation. This tome contained descriptions, including photographs, and information on associated medical issues and the genetic heritability of nearly every syndrome seen in humans, from the most common such as Down syndrome to less common ones such as Apert syndrome, to the extremely rare such as Cornelia de Lange syndrome. More often than not, the team would come across a patient with a disorder with which we were not familiar. The lot of us would gather over the dog-eared pages of this textbook to try to find what the patient had, based on the physical findings and genetic testing. I still have my own copy of *Gorlin's* in my office, but I haven't cracked it in years. Nowadays, if a child comes in with an unknown disorder, we, just as my friend the movie director, can tap in descriptions of facial features, genetic testing, and other parts of the physical exam, and in less time than one would have taken to thumb through an index, our laptop, desktop, or even phone will spit out a nifty list of possible diagnoses.

Resident teaching conferences in past years used to involve harsh grilling sessions, where we attendings would present a patient to the group, pick the most anxious-looking trainee, and expect the trainee to recite the most current literature from the journal articles he or she had copied that morning. In the hour before these conferences, the residents would huddle in a small office, flipping through pages of articles they had collectively found that morning or the evening before. Nowadays, before we even finish asking the question, residents are using just two thumbs to type in the query on their smartphone, and a list of possible article references from sites such as PubMed or Google Scholar pops up in a heartbeat. While we no longer get to see the trainees sweat it out, this remarkable improvement in access to information has changed medical and surgical training worldwide. Not only does one have access to the latest research articles, one can see videos of surgeries, of diagnostic studies being performed, and even of patient testimonials. The downside is that little if any thought is required in searching for information, and while academic institutions may have more ready access to scholarly publications, we as doctors primarily see the same sites and articles seen by nonmedical Web searchers. While we do have access, via our university or hospital passwords, to more academic sites such as PubMed and Google Scholar, we often see the same articles that our patients can see.

Not only are we looking up new information online, we are looking at patient information nearly exclusively online. The days of paper charting and file rooms are history. Electronic medical record systems are now used for an overwhelming majority of patients, initially meant to streamline and unify care. In many ways, they have. In my institution, I can find information about other physician visits, I can view laboratory test results, X-rays, and operative reports. In the near future we will be able to link this to other hospital systems, to provide an even wider net of coordinated patient care. Patients can also communicate with doctors via such record sites, via an in-network emailing system, and they can securely access their own records, without having to request reams of paper charting from a medical records department. All sounds great, right? Well, nothing is perfect.

Physicians are now even more locked in to charting, with the cumbersome requirements to click boxes and fill in fields that are, for the most part, completely irrelevant to patient care. Electronic medical record-keeping has led to increased charting time, decreased direct patient interaction (doctors now look at screens, not patients), and to its share of errors. In the heyday of the Ebola virus epidemic of 2014, when it had entered the United States, a man who had recently been to Africa came to an emergency room in Texas with fever and fatigue. The electronic medical record system had autopopulated his recent travel history after he mentioned it to his initial caregiver, but then a provider taking over didn't notice this critical piece of information. The man was sent home from the emergency room with a diagnosis of a mild flu-like illness, but he had an active Ebola virus infection. Computers can take over a lot of work, but not human-to-human contact. Had the providers communicated directly, as opposed to relying on the clicks on a screen, the man's travel history would likely have set off concerns for his diagnosis. Our increasing use of automated health-care information is growing as exponentially as our decreasing use of human-to-human communication.

WHY IT'S SO HARD TO CHANGE YOUR MIND

Nothing in medicine is black-and-white, but there's plenty of sound data—if you can find it—to help you make the best decisions. The problem is, many people choose the wrong options based on misinformation, or they continue in erroneous beliefs and biased opinions, accepted as dogma, despite irrefutable evidence to the contrary.

The explosion of widely available health-related information leads to misinterpretation, and Americans more easily trust people with high profiles, fancy websites, lots of likes and followers on social media, and stories that tug at our heartstrings than dry medical data with no real people attached to the graphs on a page. Joseph Stalin, Russia's notorious dictator in the first half of the twentieth century, once said, "The death of a man is a tragedy; the death of a million is a statistic." People are more apt to trust doctors with shiny ads, oftentimes pseudo-accolades, and posh waiting rooms than those with yellowed, tattered lab coats, tacky artwork in the hallways, and receptionists who are neither models nor wannabe actors. Websites that look nice may be more appealing than those without bells and whistles that provide more solid, albeit bland, information.

While my office is an academic one, we have some pretty nice chairs, a wide-screen TV in the waiting room, and state-of-the-art technology. We're a well-dressed group, and our lab coats are clean. But don't be fooled by appearances. We are academicians who aim to provide evidence-based care. We don't just pose for glitzy websites—we write the peer-reviewed articles, we review the peer-reviewed articles, deeming them worthy or not worthy of publication, and we sit on editorial boards of peer-reviewed journals. We've sweated in labs, given licensing exams for board certification, and treat all comers in our emergency rooms. When we're not in the office with new floors and laser machines, many of us are working in county facilities and VAs. That being said, many of us, myself included, have individual websites. My site is informational—I do not advertise my practice per se. I do, however, promote books and media events. Thanks to the power of the internet, I can track when people are looking at my website, what their landing page is, how long they stay on the site, what city they are in, their search engine, etc. Many are viewing my website while in my waiting room. How do patients know that the information on my site is reliable, accurate, up-to-date, and unbiased? They don't. But they've come in to see me, so they likely have some preconceived notion that what I say must be right. Right?

When patients entrust their care to a given doctor, deciding that doctor is trustworthy, it links to an ever-increasing challenge in health circles that is adding to the confusion: the "Curse of the Original Belief."[1] Once you believe in something, it's hard to change your mind, despite evidence to the contrary. This can be harmful to individual and public health, as well as divert people's limited attention from real dangers. If they like a doctor, they will like their website, and vice versa. I've heard countless conversations along the lines of "That doctor charges [X] dollars, just for a consultation. She must be the best." Preconceived notions of most anything—a place, a culture, a religion, a gender, or, in this case, good medical care—are hard to change. As open-minded as you might think you are (or want and strive to be), internal bias is real, and not necessarily a sign of overt discrimination. This has been studied on countless occasions in training human-resource professionals how to recognize, and ideally avoid, bias. But this takes conscious effort. The unconscious bias related to what you *think* should be good more often than not overtakes your ability to assess a situation from scratch.

THE GOOD AND BAD SIDE OF GOOGLE (AND OTHER SEARCH ENGINES)

Perhaps more noteworthy than any individual website, newspaper, television show, or magazine is the behemoth Google. This amazing power has become so widespread worldwide that, in 2006, both *The Oxford English Dictionary* and *Merriam-Webster's Dictionary* added *google* as a verb to that year's print and online editions. And you can google that. Don't get me wrong: Google is an amazing entity. I use it all the time—all doctors do—and one of its offshoots, Google Scholar (www .scholar.google.com), is a bona fide academic search engine. Google Scholar's titles are listed both chronologically and by topic. The list order is also determined by the number of citations a given article has received in academic publications. In academic circles, the number of times a given article is cited in other academic articles is one metric to measure its quality, validity, reproducibility, and even popularity.

Google has many strengths, yet some of these strengths, such as rapidity of high-volume information, order of information provided, and how that order is determined, can be quite skewed, especially when it comes to searching medical information. Let's try googling something commonly searched: *breast cancer treatment*. No surprise that it's frequently searched given that women have a one-in-eight lifetime risk of getting breast cancer (meaning that for every eight women in the United States, one will be diagnosed with breast cancer during her lifetime). When I googled this term, it took 0.52 seconds to come up with 110 million sites (the dynamic nature of the internet means you may find a different number of sites). Enough information for you? Should keep you busy while in the doctor's waiting room. But look a little closer. The top four sites are advertisements (you will load pages that are different from what I see today, but the gist of this lesson remains the same). That green ad box should be an alert that the site is biased, or at least commercial in nature. This does not necessarily mean that it is bad, inaccurate, or trying to swindle you, but it is what it is— an advertisement. The American Cancer Society's page doesn't appear until number five. At the bottom of Google page one are three more ads. Page two is similarly laid out—four ads on the top and three on the bottom of the page. Page three is no different. You get the point. These are what draw the eye in. They are the first sites you see, and the last, giving you one more chance to click on a visually pleasing ad before going to page two. Despite the 109-million-plus other sites you have access to, few people will get beyond two pages of a Google search, even though close to half of those sites presented are advertisements.

These Google ads, or AdWords, are a great way for a business to grow. The business has an incentive to use Google AdWords because the payment system is tiered; the business only pays when the ad is clicked or a visitor links to the business from the ad on the Google page. It is a slightly less blatant way to advertise, as payment is only made when the ad sees results, translated as site traffic. The larger issue than advertisements is SEO, or search engine optimization. In this free technique sites use specific key words to drive internet traffic/visitors to their site. This technique will bring in more users and will, in turn, get the site higher and higher on page one of a Google search. Because it is not inherently seen as advertising, it can be thought of as more genuine, informational, and unbiased. In general, this technique is considered legal and, more important, useful. For instance, if you are searching for symptoms of the flu, a site with key words such as *fever, body aches, cough, fatigue,* and *flu shot* will likely pop up higher than a site that just has *fever* as a key word. The gray area comes when SEO becomes SEM, or search engine manipulation. This is, indeed, a domain of vague terms. With SEM, a site overuses, or improperly uses, key words, over-stuffing Web pages, and even creating false, or spammed, links. While this practice has led to lawsuits and penalties, it is hard to regulate the fine line between SEO and SEM.[2]

So how does one possibly assess the validity of this glut of information on such an important issue as breast cancer treatment? Some basics to know are, first, whether the site is an ad. Again, it doesn't mean it's wrong, but it does mean it's biased, by virtue of being paid for by the site to Google if you click on the link. Second, look at those three letters after the dot: *.com* is commercial; *.gov* is government funded; *.org* is a nonprofit organization; *.net* is internet based, similar to *.com; .edu* is an educational site, usually tied to a university, private or public. There are a multitude of other codes, but those are the biggies. The number one site on page one of Google for breast cancer treatment is a .com pharmaceutical-company site, advertising a particular drug for metastatic breast cancer. While it also provides information, the first visual you see is the company's name, the logo, and the name of the drug.

Google recently developed a more tongue-in-cheek reputation when the actress Jenny McCarthy claimed that her knowledge of the vaccine/autism connection was obtained at the University of Google. She is not only an actress, mom, and strong advocate for her son, but also a comedian. While she said this partly in jest, she wasn't kidding. This comment went viral, providing more and more fodder for those who found her claims to be unfounded. How can one possibly consider Google to provide such university-level authority on major health issues? Let's check again. When I google *vaccine/autism,* I am provided with a mere 1,050,000 sites in 0.54 seconds. Thankfully, the *breast cancer treatment* search beats out *vaccine/autism* by about 108 million. The format is the same—four ads on the top of page one, three ads on the bottom, and the same goes for page two and beyond. As this was and remains a controversial issue, some of the sites are relaying information about how vaccines cause autism, and other sites are relaying information on how they don't. Let's assume for a moment that you have been living on another planet and have never heard or read anything about this controversial issue. The overwhelming extremes of differing information, both appearing quite dogmatic and true, are startling, if not confusing.

It is refreshing to see that not all Google searches, even those on issues fraught with controversy, have the same format of ads. When I search *cell phones/brain cancer,* another contentious issue both in the medical and the lay communities, neither ads for purchasing iPhones nor ads for enlisting a brain surgeon to remove an iPhone-induced tumor pop up first. On the contrary, in 0.52 seconds, I find 6,170,000 sites, the first of which is from the National Cancer Institute, which refutes the notion that cell phone use is linked to a higher incidence of brain cancer. One advertisement is on the bottom of page one. Alas, page two is nearly half advertisements, framing the few non-ad sites in the center of the page, but perhaps page one would have provided enough information on the issue.

Let's try a debate that is not well-known in the press, but one that is discussed in medical communities: microwave radiation exposure in utero. When a pregnant mother uses a microwave, is she predisposing her unborn child to a higher risk of developing a cleft palate, whereby the upper lip and roof of the mouth fail to come together prior to birth, leaving a gap in the middle of the face? While these children are, for the most part, otherwise healthy, they endure many major and minor surgeries during childhood, beginning at age three months, all the way into their late teens. The use of prenatal folic acid supplements has been found to substantially decrease the risk of this anomaly; however, a large genetic component remains, having little or nothing to do with maternal exposures to microwaves or use of supplements. That said, several groups believe a mother's exposure to microwaves from those ubiquitous kitchen appliances can increase the risk of having a child born with a cleft.

When I google *microwave exposure cleft palate,* a mere 948,000 results appear in 0.59 seconds. Clearly not a hot-button issue. The freelance writer looking to publish the next *US* magazine exposé should best look elsewhere. Google gives no ads at the top of page one—on the contrary, the first eleven sites are academic research articles, many of which involve laboratories and mice. Three ads are at the bottom of the page—two for nonprofit organizations that treat children with cleft palates around the world (for free), and one for a chemical company that provides microwave digestion for chemical compounds. Hardly glitzy. There are neither ads to sell microwave ovens nor to ban microwave ovens in efforts to stave off cleft palates. Perhaps one day this will become a talk-show-worthy hot topic. But for now, the Google search remains pure, untainted, and what many would consider to be boring: It's just information.

Google is not alone in its interesting search findings. If you study other search engines, they will follow similar formulas. So how can you possibly navigate important issues when there is so much biased information? A few simple tools to remember:

1. Be aware of the ads. While many of these have useful information, remember what they are: advertisements. They are selling something—a product, a treatment center, a diagnostic test, or even a nonprofit asking for donations. They may simply be wanting to drive up traffic to their site, but they are paying to do so.
2. Look for .org or .gov sites first. While many of you might be wary of government-funded sites, they do provide unbiased data from larger studies.
3. Be wary if a pharmaceutical company is the site provider. While many drugs are well studied and are lifesaving, hold off on these sites as your first go-to on learning about an illness or a condition.

These tools can apply to any condition, symptom, or health question you may have. Chances are, the less controversial, newsworthy, or even common the condition, the more informational sites over advertisements will appear.

We can't blame the internet for our flaws in falling prey to hype and mythology in medicine. More than a hundred years ago Clark Stanley made himself famous when he gathered a large crowd and pitched the miracles of snake venom. He would kill rattlesnakes while touting the benefits of the reptiles' secretions, claiming his medicinal concoctions were derived from a secret recipe of an Indian medicine man. His miracle juice, or oil, would cure all manner of ills, including toothaches, sprains, and general pain. Fifty cents a bottle could buy your antidote (which is the equivalent of about $13 today—not bad). In 1917, federal agents decided to see what was in Stanley's medicinal wares. It was indeed a type of snake oil at 99 percent mineral oil, a bit of beef fat, and some red pepper and turpentine to give it that quintessential taste of medicine. Stanley was soon out of business, but the term *snake oil salesman* lived on and thrives today in all forms of fakeness that go far beyond just health circles.

SHOULD YOU YELP YOUR DOCTOR?

What about doctors, hospitals, and treatment centers? Many of us go to the website of the caregiver or hospital. If the site is clean, user-friendly, easy to navigate, and provides some useful information, that's often a win. If the site is filled with product placements, sales pitches, and exaggerated claims of unrealistic outcomes, raise an eyebrow. But what about the rating sites for doctors? My institution has an internal rating system, provided confidentially to the doctors, filled out by patients. If we get great ratings, we love the system. If our ratings fall, the system is clearly flawed. I find my ratings to be pretty useless, by virtue of the simple fact that so few patients are asked to participate. On average, such an extremely small number of patients fill out the ratings form, but I somehow receive a complicated calculated score, a percentile rank compared to those in my field, and a few other data points, which may not have any bearing on actual quality of care. As few as five or ten patients out of the thousands I see in a twelve-month period may determine my score.

And what about the public rating sites? Who is posting reviews of doctors, dentists, hospitals, and clinics, and are these reviews valid? Yelp is a popular one of these sites. It is a free site and app that requires you to sign in to get access to much of the information. It was established in 2004, with the goal of "help[ing] people find great local businesses like dentists, hair stylists and mechanics." They have an average of 26 million users per month on the app, and more than 80 million users per month on the website. As of late 2016, 127 million reviews were posted on the site, more than forty-two thousand of which described U.S. hospital experiences.[3] It is transparent in that it states that it makes its money from advertisements. It also states that businesses may not change the reviews they receive.

Can you safely use a site such as Yelp to glean information on your doctor, hospital, or dentist? The answer is complicated, as you do not know the caliber nor the honesty of the reviewers. Online reviewers tend to post a review if their experience was either extremely positive or extremely negative. If a patient had a terrible time finding parking for an appointment, arrived late, and had to wait an extra half hour to see the doctor, he may give a review along the lines of "waited too long to see the doctor." If a patient finds a great parking spot, arrives on time, and gets seen in a timely manner, she will likely provide a different review. Neither of these reviews has any bearing on the care the patient received, but both remark on the overall experience. These same patients also likely checked out the other reviews prior to posting their own. It is human nature to want to be in line with others and to write a review similar to what is already seen. This goes back to the Curse of the Original Belief: If you think the care is supposed to be excellent, you will be hard-pressed to say—and believe—otherwise. If you walk in a skeptic because you've read some bad reviews, you will find every reason to write a bad review as well.

I think Yelp is a great source for reviews of businesses such as restaurants, hotels, and vacation spots. There are usually multiple reviews, and many of them are entertaining. I'd love to know that the pasta

Parmesan made someone sick while the following reviewer orders it every time she goes and never had anything but an excellent meal. But doctors are different. The outcome is not so simple. If a patient has a complicated illness that has no viable treatment, is the doctor bad? If a patient has an easily treatable illness that the doctor misses, but he saw the patient on time, had a courteous staff, the patient recovered despite inappropriate care, and the office validated parking, does that make the doctor good? Yelp reviews might say so. Sites that enable such quick reviews can be misleading, to say the least.

There are ways to rate doctors and hospitals more objectively, although even these have several flaws. Unfortunately, these sites are less readily accessible and less on people's radar than Yelp. The Leapfrog Group was established in 1998 as a hospital-rating watchdog organization that provides detailed data on complication rates, successful and unsuccessful outcomes of multiple treatments, numbers of unscheduled readmissions, and so on. Based on voluntarily supplied information from hospitals of specific outcomes of specific procedures, analyses of metrics can be made to improve patient safety and minimize poor outcomes. For example, hospitals where more than five hundred coronary artery bypass graftings (CABG, pronounced "cabbage"— open-heart surgery to repair blocked arteries) are performed annually fare better in rankings than those where fewer than five hundred procedures per year are performed. This is also true of hospitals performing abdominal aortic aneurysm repairs and premature deliveries.[4] Some assessments of the accuracy of this method of quality measurement have found that these numbers do not necessarily hold true, that more cases do not necessarily equal better outcomes.[5]

One bias of Leapfrog data is that it does not clearly account for differing patient populations. For instance, a hospital that performs heart surgery on patients with no other medical complications, such as prior unsuccessful heart surgery, preexisting medical problems such as diabetes, lung disease, vascular disease, or advanced age, will likely have better outcomes and Leapfrog data than a hospital that handles heart patients with highly complex conditions.

Everyone wants her doctor to be a superdoctor. But what does it mean to be a Super Doctor? Super Doctors is a group that describes itself as a "listing of outstanding physicians . . . who have attained a high degree of peer recognition or professional achievement." The selection requires peer nomination, with efforts to "prevent ballot manipulation, and doctors cannot self-nominate." Doctors who receive the highest scores are invited to review other nominees in their field. Those considered to be highly qualified in their fields, based on peer review, are invited to be featured in a Super Doctors issue of a widely distributed local magazine. Doctors are evaluated by peers, not by patients or outside institutions. The selection is based on such features as years of experience, professional awards, publications, leadership, and board certification. So far, all sounds legitimate. And for the most part it is. I am a Super Doctor! But the kicker comes later. If you want to be a featured Super Doctor, you are then strongly invited, on multiple occasions, to buy an advertisement in their publication feature. This payment may be to simply make your name bigger and bolder, or it can be up to a full-page color advertisement for your practice. Peer-reviewed notoriety or ad? Ad.

Castle Connolly is a similar peer-reviewed "top doctor" site. Their selection process is similar to that of Super Doctors, with doctors nominated and ranked by peers in their field. On their site, the Top Doctors Resources link brings us to the option of advertising in *The New York Times*. Yes, it's a peer-reviewed process at Castle Connolly, but it's tainted by the lure of self-promoting with payment. Advertising. Let's say you need to find a doctor but you're not sure who's the best. Let's use Google and find the "best cardiologist in Los Angeles." The first link that appears is Castle Connolly. The second is Yelp.

HEALTH WEBSITES AND THE ART OF SELF-DIAGNOSING

If you are already immersed in looking up health issues, be they symptoms, medications, illnesses, hospitals, or doctors, you are likely familiar with many of the numerous well-established health information sites. They abound, and you may be versed enough in health sites to bypass such search engines as Google or Yahoo! Some sites claim to be (and are) purely informational, while others are commercial, with multiple advertisements, either embedded in the text, as pop-up ads, or as advertisements by website sponsors. When examining and utilizing health websites, keep in mind the many types, including those offering general health information, hospital-based websites, insurance company websites, disease-specific websites, pharmaceutical company websites, noncommercial websites offering drug information, and websites offering information on physician credentials. Given the lack of vetting in the launch and maintenance of a website, you have several tools, and some basic information, to allow you to navigate this entangled web of easily accessible health information.

The many general health websites tend to be large and popular and cover a wide range of health information. In many cases, the content is provided by or edited by physicians. Among these general health sites are commercial ones (e.g., www.webmd.com), nonprofits (e.g., www.mayoclinic.org), and federal-government-run sites (e.g., cdc.gov). The commercial sites have substantial advertising. This could introduce bias, but these websites routinely disclaim such influence. The nonprofit websites have little or no commercial advertising, but they do promote their own services. The federal government websites offer what could be considered unbiased information, with no advertising, and emphasis on information extracted from large, evidence-based studies. Some websites also offer a "symptom checker." It's easy to fall into the trap of self-diagnosis (or diagnosing family and friends) based on these easy-to-navigate symptom checkers, but keep in mind that a list of symptoms does not equal a diagnosis. Symptoms of a migraine headache may be similar to those of a stroke. Symptoms of a heart attack may be similar to those of a stomach virus or sore shoulder. Symptom checkers should only be used to guide you to seek a healthcare professional's evaluation, not to make the diagnosis or treatment plan on your own.[6]

In general, searching online for health information can be valid, eye-opening, educational, and even useful. While many practitioners roll their eyes when they hear, "I did my research," from a patient, sometimes that research can be sound. If a patient has a rare disease and presents articles about it, many of us will be grateful that we were saved some extra work. But the Web becomes entangled when sites angled with opinions, personal anecdotes, blatant exaggeration, and false claims manipulate the navigator to believe what is posted. People also run into trouble when looking for information online based on preconceived notions. Here comes the Curse of the Original Belief again. If you believe that megadosing on vitamin C will prevent colds, you will seek out (and easily find) sites promoting this notion. If you think that juice cleanses are the way to better health and well-being, it's easy to find websites supporting this. If delaying vaccines is your cup of tea, online sources abound (but please read Chapter 10 first). If you're debating whether to eat only organic food, plenty of available information will support this.

In the deep waters of the internet's sea of health information, swim carefully. When searching, use terms that do not have an opinion embedded in them. Instead of *megadosing vitamin C to prevent colds,* start with *vitamin C* as your search. Learn how this vitamin is absorbed, how it's metabolized, what a deficiency looks like, what an overdose looks like, and what it looks like to have just the right amount. Always start with the basics. Take time to educate yourself about the nuts and bolts before digging into the extremes or controversies. If you're concerned about vaccine risks, learn about how vaccines work, what diseases they prevent, and what their benefits are; then you can dive into potential risks and complications. Health information online will only continue to grow. Your own health information is now online. Most doctors' offices and hospitals have electronic medical records. How do you want your health-care providers to search your information? It should be by fact, not opinion.

If the name John Harvey doesn't sound familiar to you, that's probably a good thing. I won't tell you his last name until after I describe who this man was and what he did. John Harvey was born in 1852 and grew up to become a medical doctor in Battle Creek, Michigan. He and his wife never had biological children of their own but eventually adopted eight of the forty-two kids they fostered. Legend has it he was a strange character. John Harvey was famous for a sanatorium that he ran in Battle Creek, where wealthy individuals went to get healthy using holistic methods. The sanatorium, what we'd likely today call a medical spa or longevity center, promoted high-fiber vegetarian diets, regular exercise, no smoking or drinking, and deep-breathing exercises—all positive actions for health. But some of its other practices we might question today. John Harvey was a huge fan of enemas, for example. He prescribed them regularly to his patients, who would receive several gallons of water up their intestines from a machine that forced the flow. He believed sex weakened the body, so that was banned. All sexual activity was prohibited, including masturbation, which he thought caused cancer of the womb, urinary disease, and epilepsy, among other illnesses. To reduce sexual pleasure, he prescribed circumcision for men and the application of carbolic acid on women's clitorises. Caffeine was considered poisonous, so coffee and tea were not allowed. This is a small slice of what John Harvey believed and endorsed. Although his treatments were extreme, he had quite a following. Among the people who took to his teachings and practices at the sanatorium were former president William Howard Taft, Amelia Earhart, Henry Ford, and Thomas Edison.

Now you may be wondering what else John Harvey was famous for. His last name: Kellogg. J. H. Kellogg is perhaps best known as the inventor of corn flakes, the breakfast cereal, a brand of which bears his name. He and his brother Will K. Kellogg patented the corn flakes process in 1896. By 1928, Kellogg's had started to manufacture Rice Krispies. Makes you wonder what John Harvey would think about Rice Krispies treats today given that he's considered the father of the modern health food movement.

How many stars would J. H. Kellogg's longevity center get today? Would you see good reviews on Yelp? I'm not so sure.

HYPE ALERT

* Major search engines such as Google, Bing, and Yahoo! provide a wealth of information, but be vigilant of ads, overblown claims of success, and exaggerated testimonials.

* Learn to navigate health questions online. Be as specific as possible. Big, vague questions give you inordinate options for answers. It is easy to get bombarded by these.

* Most online reviews of doctors are from extremely satisfied or extremely dissatisfied patients—rarely from those in the middle of the road.

* Don't Yelp your doctor! A large movement is afoot in the medical field to block this.

* Just because you read something on a glitzy website doesn't mean it's true.

2

RISKY BUSINESS: WHAT EBOLA AND YOUR CAR HAVE IN COMMON

How to Put Risk into Perspective

SHOULD YOU WORRY ABOUT EXOTIC GERMS SUCH AS
ZIKA AND EBOLA?

WHAT IS THE NUMBER ONE CAUSE OF DEATH BEFORE
THE AGE OF FORTY-FOUR?

WHAT IS MORE POWERFUL, YOUR GENES OR YOUR ENVIRONMENT?

IS IN-HOME GENETIC TESTING WORTH IT?

In the fall of 2014, the news headlines had millions of Americans glued to the media in hysterics about Ebola, a virus that has a high mortality rate and had landed in the United States from West Africa, where it had killed scores of people. People started to ask questions such as whether they could get Ebola from a toilet seat, in a stuffy subway car, or from a dog exposed to it. We Americans do panic well, and we seem to love worrying about exotic matters rather than ordinary, familiar ones, which we treat rather cavalierly.

A few cases in point: According to the Centers for Disease Control and Prevention, less than half of us get the annual flu vaccine, even though each year in the United States the flu kills up to twenty thousand people and lands over one hundred thousand in the hospital, and the vaccine can typically prevent or at least soften the blow of the virus if it's contracted. In a bad year, when the flu is especially virulent, up to sixty thousand people in the United States will die if they are unvaccinated.[1] Tens of thousands of Americans perish in car crashes each year, and more than half of those people weren't wearing seat belts. Nearly a quarter of teenagers in fatal accidents are distracted by their cell phones; every day eleven teenagers die as a result of texting while driving (car crashes are the leading cause of death of teens in the United States).[2] And vanity must trump sanity when it comes to tanning: more than 3.5 million individuals are diagnosed with skin cancer yearly and nearly ten thousand of them die.[3] Today one in five deaths in the United States is now associated with obesity.[4] Over the two-year period of the Ebola virus "outbreak," there was one U.S. death. So, indeed, éclairs are scarier than Ebola.

ACRONYM ANGST

Few remember the acronyms GRID, or HTLV. In the early 1980s, a mysterious constellation of acute and deadly skin cancers and pneumonias in previously healthy men in several large U.S. cities occurred without explanation. Case reports were presented in esoteric medical journals. HTLV-3, or human T-cell lymphoma virus, led many medical specialists to scratch their heads in puzzlement, until those at risk became clear: the illness was seen exclusively in previously healthy homosexual men. These clusters of affected gay men led to the term GRID, or gay-related immune deficiency. This soon became known as human immunodeficiency virus (HIV), expanding the potential involvement to all humans, not just men who have sex with other men. So began what became known as the AIDS (Acquired Immunodeficiency Syndrome) crisis, and eventual epidemic.

AIDS was the talk of many towns throughout the world for several decades. People with AIDS were shunned—lost their jobs, found it hard to get medical care, and were banned from many schools, sports teams, and public places. As it initially affected gay men, intravenous drug users, and those from several African countries, the social stigma tied to the illness became perhaps more newsworthy than the illness itself. Unlike other sexually transmitted diseases such as chlamydia, gonorrhea, or syphilis, HIV is much more readily transmitted via blood. Anal intercourse is associated with more skin breakdown and blood exposure in the anus or rectum, even in microscopic amounts, making HIV transmission via anal intercourse eighteen times more likely than during vaginal intercourse. In the early days of HIV, well before it was recognized as a sexually transmittable disease, gay men were less inclined to wear condoms during sexual encounters than they were when HIV became more widely recognized as a virus transmitted during sex. Healthy folks feared that shaking hands with a person with AIDS would lead to their contracting the disease. Many doctors feared contamination from treating a patient with AIDS. I do not know a surgical resident who, when I was training in the early 1990s, didn't panic at a needle stick. Many of us would not only get tested for HIV, but would take AIDS treatment regimens for several months, in efforts to prevent the virus from entering our system after a potentially HIV-contaminated needle stick. While AIDS is still a tragic illness, nowadays many more live with HIV infection than die from it, thanks to medical therapies (namely drugs) that help keep the virus at bay. And its transmission has been dramatically reduced. If a woman with HIV becomes pregnant, transmission from her to her unborn baby is nearly 100 percent preventable with proper treatment. AIDS still exists and still kills, but it's out of the spotlight. It's not on people's radar, it's not headline news, so we don't panic about it. Worldwide in 2014, there were over twenty thousand new cases of Ebola, less than a handful of which were in the United States; there were over 2 million new cases of AIDS in that same year.[5]

As we are so tied to the twenty-four-hour news cycle, the sensationalized devastations such as previously unheard-of deadly diseases, terrorist attacks, school shootings, and natural disasters tend to bring out our mortal fears and push our panic buttons far more than the chronic, unsexy risks of being overweight, not getting enough sleep, not wearing a bicycle helmet, or not staying up on our vaccinations. Perhaps the traumas and tragedies over which we have little or no control draw us in. As when watching a horror movie scene, where we close one eye but keep one open to gape at the gore, deep down we want to see the horror and even partake in the shared shock. Risks of wild viruses, which now include Zika, not just Ebola or HIV, bombing attacks and shootings, and succumbing to natural disasters are always present, but they are close to impossible to prepare for and can be hard to prevent. We can practice safe sex, avoid direct contact with an acutely ill person, and avoid mosquito-infested lands, but there will always be another virus after HIV, Ebola, and Zika.

We can take precautions in schools and crowds, but mass destruction is usually a surprise—despite all our best efforts to prevent or recover quickly from it. We can have an earthquake survival kit, an extra water supply for extreme drought, sandbags to stem rising tides during a flood, or an in-home generator in case of a hurricane-induced power outage, but disasters will nonetheless occur, and our seeming preparedness may be for naught if the disaster crushes our emergency supplies. One can do only so much preparation for natural or human disasters. But in earthquake-prone locations there are likely more folks who have earthquake kits than those who watch what they eat, break a sweat regularly, or stay off their phone while driving. We're all waiting for the next disaster, but who's bracing for a twenty-year progression of cardiovascular disease?

New Year's Eve 1999, Y2K, as we called it, was one of the most hyped, anticlimactic moments ever, as all of the exciting, emotionally driven disaster preparation turned out to be a waste. People were expecting the world to somehow end as the Y2K clocks would fail to strike midnight, as computer systems, timed alarms, businesses, power grids, ATMs, and so much more would somehow not be able to recognize the end of one millennium and the beginning of another. Back in the dark ages of computer science (the turn of the millennium), even the most sophisticated computer programs and their brilliant human programmers were unable to predict the outcome of the turn of the clock that night. Nobody knew how to prepare, so as we Americans are wont to do, we did everything: stocked fridges, cleared grocery stores of water jugs, filled gas tanks, backed up computers as never before, cleared out bank accounts, built bunkers, and hid. Many hid with a year's supply of toilet paper, which would be much needed after eating and drinking their well-stocked food and water supplies. As luck (or logic) would have it, the clocks struck, the computers worked, the lights stayed on, and life went on as usual on January 1, 2000. Most breathed a sigh of relief, many laughed at themselves for overpreparing, but some were disappointed at the lack of Sturm und Drang for the well-planned-for apocalypse that wasn't. Less newsworthy was that the peak of the annual flu season was between Christmas 1999 and the first weeks of January 2000. Many of those illnesses could easily have been prevented with a flu shot instead of a stocked pantry.[6]

I deal with risk daily in my work. I help families decide whether surgery, watchful waiting, or medicine is the best treatment option. Most would think that surgery carries the highest risk, is the most aggressive option, and is a last resort. This is absolutely not the case. Yes, surgery has risks—from the anesthesia, the procedure itself, and short- and long-term complications afterward. I spend most of the counseling time going over these risks. What's the chance of an allergic reaction to anesthesia? What's the chance that I won't wake up? Will there be excessive bleeding? Will the recovery be miserable, fraught with complications? Will the surgery help? We have answers, given as percents or thousandths of percents, for the patient. But if it happens to you, or your family member, the risk is 100 percent, and if it doesn't happen, the risk is zero. While this logic is illogical, the risk concern goes out the window when the presented risk does or does not happen. What people don't readily consider are the risks of no surgery, or no medical treatment.

ABSOLUTE RISK

All treatments, including surgery, bear risks, but the benefits of intervention may outweigh the risks of no intervention. Obstructive sleep apnea (OSA), for example, is a common disorder in which the airway collapses during sleep as the muscles in the back of the throat fail to keep the airway open. One's breathing gets cut off multiple times and sleep becomes fragmented. Dreamless sleep and loud snoring are its tell-tale signs. The potential risks are sleep deprivation, poor growth in children, attentional issues with misdiagnosis of ADHD, hypertension, heart abnormalities, and, in adults, stroke. When a child is being evaluated for ADHD, a thorough history of sleep issues is taken to assess whether the ADHD symptoms either originate from or are exacerbated by a sleep disorder. About 3 percent of children have a sleep disorder, and about 10 percent are diagnosed with ADHD, so there is likely some overlap in these diagnoses. If you have OSA and receive no treatment, risks of these related illnesses increase. Watchful waiting in the case of a child can be even more harrowing than in an adult. Countless parents have brought their children in to see me over the years when the "cute snoring" of young childhood gradually turns to more worrisome night-time breathing sounds. Many are told not to worry, that this noisy breathing is simply a rite of passage, that their child will grow out of it or that it's just a phase. In some, this is absolutely the case, but in others, the child is actually suffering from obstructive sleep apnea, where watchful waiting is most definitely not the way to go.

When the parents of a six-year-old boy started to get calls from the child's teacher regarding disruptive behaviors, inattention, poor focus, and even some difficulties in speech and articulation, the astute teacher asked a simple question: "How's he sleeping?" His mother replied, "The same as always—he rocks the house. We know he's asleep because we can hear his rumbling breaths, sometimes his coughs, and later in the night, he wakes up and comes into our room." As the parents heard what they were saying, they knew this wasn't right.

I saw this boy, whom I'd describe as skinny and squirmy, trying to stay still in my office exam chair. He didn't look sleepy, as you would classically picture an adult with poor sleep. This kid was tired and wired. Also noticeable was the way he sounded. He breathed like Darth Vader and spoke as if he had golf balls in the back of his throat. Well, it wasn't quite golf balls, but was something pretty similar. Large tonsils, each the size of a hefty meatball, were filling the back of his mouth. His adenoids, tonsil-like tissue that sits in the back of his nose, were equally large.

When I discussed his sleep habits, as well as his daytime issues, with his parents, such as irritability out of proportion for his age, difficulty with paying attention at school and at home, dietary pickiness, speech sounding as if he always had a cold, and chronic breathing through the mouth instead of the nose, it was clear that he needed to get some sleep, some good sleep. Even though he was in bed for the requisite ten hours each night, his noisy, obstructed breathing was affecting his sleep cycles, especially the deeper stages of sleep, including the periods of rapid eye movement, or REM, sleep. He was effectively living on five or six hours of sleep per night, far below what his body and brain needed to function properly. I'm cranky when I'm sleep deprived, and I'm sure he was, too.

The following week, he came in to have his tonsils and adenoids removed. I'll be honest—the surgery is no picnic for these kids, nor for their families. But with improved anesthetics and surgical techniques, the recovery is not as horrendous as it used to be. The surgery took a half hour, and two hours after that, he was home, eating noodles. But what's more remarkable was the call I received from his mom three weeks later. She was in tears. "I can't hear him! I check on him all night because his breathing is silent. He sleeps all night, is the first awake in the morning, and his teacher can already notice a change in his mood, focus, and behavior."

I told the mother to stop crying! (No, I didn't—I'm not *that* heartless.) But she was certainly not the first, second, or third parent I've heard that from. This child was at risk for longtime school issues, poor performance and self-esteem, exhaustive behavioral interventions, and speech therapy. A little intervention, albeit not without risk, was life changing. But an important caveat to this happy ending: the lines do get blurred when diagnosing veritable neuropsychiatric disorders such as ADHD and/or autism spectrum disorders in children with sleep issues. Sometimes they go hand in hand, and you don't know which came first. Sometimes they act in parallel, a child having sleep apnea *and* ADHD, not one causing the other. Both are common, and one should always seek evaluations and recommendations first from one's primary care doctor, followed by advice from specialists as needed.

Sometimes causes can be a bit unclear, as is often the case in a young child with behavioral issues. Is it a phase, a disorder, or a physical problem? But sometimes the issue is direr and we need to act quickly. Very quickly. I will never forget the fifteen-month-old boy I treated a few years back. He had a loving family who wanted only the best for him and his three-year-old brother. His parents wanted to have them exposed to healthy foods, which included nuts. Every five days in this country a child dies from choking on food, but few folks know this unless they're in my business. Most of these children are under age five, and the most common choking hazard is nuts. This beautiful fifteen-month-old had a bad cough that wasn't getting better, so the parents sought advice and evaluation from their pediatrician. The pediatrician recommended a chest X-ray, but the family was concerned about risks of unnecessary radiation exposure to their toddler. They held off on the chest X-ray until the following week, when the cough was getting much worse. The chest X-ray showed that one lung was completely blocked, meaning no air was getting through to it, but the other lung looked fine.

Reflecting back, the parents remembered that the baby coughed briefly—about three weeks prior—while eating cashews. The pediatrician recommended an evaluation for surgery to remove a potential nut in the child's windpipe, blocking the lung, but the parents were extremely concerned about the risks of surgery. After several days of convincing the parents, the child did undergo surgery to remove the nut. The child almost stopped breathing during the surgery. He could have died on the table. The risks of the surgery were escalated exponentially by the choice to wait. The longer a foreign object sits in an airway, especially a food item such as a nut, which exudes oils that inflame the tissues, the more swollen the airway gets. The more swollen an airway gets, the more blocked the airway passages become. By the time this boy came to surgery, the nut had caused so much swelling and inflamed tissue that his entire right lung was blocked. Putting a child to sleep with only one functioning lung is risky. When anesthesia is given, the breathing needs to be supported by either the person's breathing on his own, albeit while unconscious, or by a mechanical ventilator. Even perfectly functioning lungs have some challenges during surgical airway procedures. When one lung isn't working, it makes for quite a harrowing time. These risks would have been avoided had the child not eaten nuts. The American Academy of Pediatrics recommends that children under age five not eat nuts of any kind, due to the high risks of their choking. Hindsight is twenty-twenty, but risk perspective is often skewed by what one sees as risky as opposed to what really is. While getting X-rays or having surgery poses risks, the risks from a young toddler eating nuts is the much higher.

Until the advent of antibiotics, starting with the discovery and development of penicillin in the first half of the twentieth century, life-threatening infections from bacteria were common throughout the world. Babies would die from bacterial meningitis; a mild case of tonsillitis could easily spread to the kidneys or heart. Sore throats would extend into the chest. Finger infections would commonly require amputation. Most did not survive brain abscesses. Many succumbed to appendicitis. Antibiotics were miracle drugs. But they have become more and more dangerous. They are now, in part, responsible for many of the illnesses they treat. They are one of the most important and now one of the most overused medications in modern medicine. The vast majority of prescribed antibiotics are unnecessary; so many illnesses are viral, not bacterial, and would resolve without medications. The overuse of antibiotics has led to a worldwide crisis of antibiotic resistance, whereby the strong bacteria have survived and even morphed into the dreaded superbugs that don't respond to any drug treatments. It has become a literal Darwinian survival of the fittest in the world of bacteria. If you've ever had a strep throat, bronchitis, a sinus infection, or even pneumonia that didn't respond to an antibiotic, it's probably because the bacteria wreaking havoc on your system has already outsmarted the run-of-the-mill antibiotic and will survive despite the antibiotic's attempt to kill the infection. Your doctor will likely either double your dose (some bacteria are partially resistant and will get knocked down with higher dosing), extend your dose, or switch antibiotics.

When I was a kid, we had penicillin, and perhaps a few other common antibiotics. But once antibiotic resistance became the norm in the late 1980s, antibiotics became an industry unto themselves. Cephalosporins (common ones include cephalexin or the brand name Keflex, cefdinir or Omnicef, and cefixime or Suprax) became the new penicillin. When I was a resident, pharmaceutical reps would lure us with gifts, meals, and more. When choosing an antibiotic, we would joke that we'd prescribe "cepha-lunch," meaning we'd give whatever drug was most recently responsible for feeding us. In my and many institutions, this activity is no longer allowed, as it was thought to create bias on the part of prescribing physicians. Indeed, it did.

Infectious-disease specialists have dedicated their careers to help minimize the extraordinary overuse of antibiotics, in efforts to allay the superbug crisis. Sadly, most of the damage has already been done, as superbugs and super-superbugs are alive and well and living in schools, gyms, day cares, hospitals, airports, and most forms of public transportation. Limiting the use of antibiotics going forward will hopefully minimize the continued evolution of untreatable microbes. An infection called MRSA, which stands for "methicillin-resistant *Staphylococcus aureus,*" used to be a serious infection seen only in critically ill, hospitalized patients. This bacteria was resistant to standard antibiotic treatments and would only rear its ugly head in someone who had previously been treated with strong antibiotics for other infections. Patients with MRSA would be quarantined for weeks or even months and treated with powerful intravenous antibiotics. Nowadays, we see healthy people walk into our offices with MRSA infections; we give them prescriptions for powerful oral antibiotics, and they go back to school or work and hopefully come back for a follow-up appointment. Superbugs are the new normal. The good news is, many are less virulent than they used to be and are easier to treat. The bad news is, we're running out of treatments and pharmaceutical companies are not spending their R&D money looking for new antibiotics.

If antibiotics are so bad and are leading to development of superbugs and risky illnesses, why aren't we doing more to stave off the temptation to give every child with a cold amoxicillin, every surgical patient unnecessary prophylaxis, and everyone with a sore throat penicillin?

Why is it that, despite clinical guidelines by many medical academies, such as the American Academy of Pediatrics, against antibiotics overuse, they are still overused? In part, it is habit (albeit bad habit) on the part of the practitioner, but part is from concern over risks of not treating. Indeed, there are real risks to not giving antibiotics, despite the latest focus on the risks of treatment. For instance, in infants and those with frail immune systems, the risks of not treating, which may lead to potentially life-threatening complications, are higher than the risks of treatment. While treatment guidelines are excellent ways for us as physicians to make clinical decisions, each patient, and each patient's specific situation, needs to be individualized.

A heavy smoker who develops a cold is at higher risk for developing bronchitis or pneumonia than a nonsmoker. So even if most colds are not bacterial, it is often wise to treat a patient with a higher risk for bacterial infection with an antibiotic. For another example, a person who has had multiple urinary-tract infections may be at higher risk for developing a kidney infection and may need stronger antibiotics than a patient who's never had a urinary-tract infection. We are humans, not robots, and our bodies have different needs at different times, sometimes regardless of published guidelines. We doctors encounter all too often calls from patients who say they have the same sore throat as last time and can we call in those antibiotics? Now we are saying no. Just as every patient is different, so is every infection. So, yes, we do need to see you. Again.

RISKY BEHAVIOR

On a busy surgical day, my team and I explain detailed risks and benefits of procedures, medications, anesthetics, and recovery as many as ten times. Imagine if every time you took a step off the curb, ate or drank something, got in your car, or hopped on a plane, you'd get a detailed list of the risks and benefits of your actions for the minutes, hours, or even months and years ahead. While this notion is ridiculous, it has some validity. You can't neglect the benefits versus the costs when you consider risks. What are the risks and costs of driving during rush hour versus the benefits? What is the cost and risk of taking the rush-hour train? Do the benefits outweigh these risks? Some of these parameters are extremely complicated if not impossible to measure, especially when you factor in lost time, stress, control of your day, and quality of life. But some decisions carry with them much clearer and calculable risks, costs, and benefits. Smoking cigarettes, for example, carries far higher costs and risks than benefits. But having a daily glass of red wine is a gray area in costs, benefits, and risk. Exercising daily will clearly come out as beneficial, but too much exercise might bring more risk than good.

How can you possibly calculate risks, benefits, and costs daily? This is no small task, but when we are confronted with so many confounding recommendations and ideas, it's always best to start big. In medicine, we have the expressions "Common things happen commonly" and "When you hear hoofbeats, think horses, not zebras." This seemingly simplistic idea applies not only to the oftentimes futile search for a rare disease masking as a common illness, but also to nonmedical folks wending their ways down pathways in search of the extreme. When it comes to our health, certain risks come with all stages of life. Since life itself is the most precious thing to sustain, let's go through the most likely causes of death in the United States for early life and then for each decade of life:[7]

1. In the first year of life, the most likely cause of mortality is congenital abnormalities, followed by premature birth, SIDS, maternal pregnancy complications, and "unintentional injury" (*accidents*)[8]
2. For ages one to five, the highest risk of death comes from *accidents* (followed by congenital anomalies)[9]
3. For ages six to fourteen, the highest mortality is from *accidents* (followed by cancer)[10]
4. For ages fifteen to twenty, *accidents* (followed by homicide)
5. For ages twenty-one to thirty, *accidents* (followed by suicide)
6. For ages thirty-one to forty-four, *accidents* (followed by cancer)

7. For ages forty-five to sixty-four, malignant neoplasms (cancer) (followed by heart disease)

8. For ages sixty-five-plus, heart disease (followed by cancer)

In sum, the leading cause of death from ages one to forty-four is accidents. The take-home message for the under-forty-five set: Be careful. Drive safely. Wear bike and ski helmets. Avoid motorcycles. Fasten that seat belt. Avoid playing drinking games or binge drinking. After that, you can start worrying about your health. Well, not quite, of course. But too few focus on injury prevention while focusing on rushing through traffic, answering that text while driving, or not wearing a seat belt.

For long-term health and longevity, factors such as safety and injury prevention are in our control. Many daily health habits are also in our control, such as a healthy diet, adequate sleep, not smoking, exercising regularly, and finding ways to reduce stress. But many factors are out of our control—our genetic makeup, the city in which we live, our job, our marital status. Many studies have demonstrated that married people live longer than single folks and suffer less depression and fewer chronic illnesses. But getting (and staying) married is no simple task, unlike eating vegetables and getting to the gym. Children who live in tight living quarters, busy cities, and near major roadways are at higher risk for asthma and lung diseases. But asking them to move is absurd. Certain professions, mine included, are associated with substantial occupational hazards, yet we continue to do what we do. Some of these risk factors, such as asbestos exposure, leading to the chronic and deadly lung disease asbestosis, or coal miner's lung, are discovered long after the damage has been done to thousands and thousands of workers.

Even more out of our control is how the future of our health can be determined before we are born—not only by genetics, but also by our environment. One of the lesser-known effects of D-day was the subsequent German ban on food supplies to parts of the Netherlands. This response to the Allied invasion left parts of the Netherlands near famine, with rations trickling down to as low as four hundred calories per day. Between the winter of 1944 into the spring of 1945, what was later known as the Dutch Famine led to a doubled rate of mortality compared to that of 1939, and widespread malnutrition. Despite such devastation, babies continued to be conceived and born. Years later, as this population grew into childhood and adulthood, scientists studied their health to find patterns in disease later in life that could be attributed to those early exposures. Those born to a mother suffering from famine had lower birth weight but higher likelihood of breathing problems, glucose intolerance, and, surprisingly, obesity. Many of these issues wouldn't appear until adulthood.[11] They even had a higher likelihood of heart disease due to coronary plaque formation. In what's now a burgeoning new branch of scientific knowledge called the developmental origins of health and disease (or DOHaD), scientists are finding that many chronic ailments usually diagnosed later in life originate from early life experiences—more so in many cases than from genetic inheritance.

On the flip side, we can control many lifestyle behaviors and medications to lower our risks of health issues, and this can start as early as pregnancy. Vitamin supplements during pregnancy, especially the B vitamin folate, lower risks of central neural defects as well as cleft palate. Smoking cigarettes during pregnancy is a known risk factor for low birth weight, premature delivery, and lung disease in infants. My college thesis looked at the association between smoking, alcohol consumption, and demographics and low birth weight in two rural Canadian towns in northern Quebec, one French speaking (the majority population, with higher socioeconomic status) and one English speaking (the minority population, with lower socioeconomic status). Low birth weight was more commonly associated with English speakers, who also smoked more and drank more alcohol than their French-speaking counterparts.

Some variations occur in the mortality figures between genders, but the general idea applies to both men and women. You may be surprised to see that some of the more likely causes of death are far from your radar. You see the numbers, know that they're real, but may be more focused on the latest superfood or power drink. You may want to be getting the requisite ten thousand daily steps logged in to your fitness tracker. Many of us, myself included, are routinely in search of that magic formula to keep us healthy, but in so doing, we lose sight of the less glamorous basics. But how many of us look at or use our cell phones while driving, disregard the speed limit, or regret something stupid we did as a teen or young adult?

In my own neighborhood, I see kids learning to ride bikes on busy streets without bike helmets, but their parents are vigilant that the children eat only organic foods and minimize screen time. The misguided risk perceptions are clear. The notion of health and safety becomes focused on branding and trends, not on real risk and danger. This is, in part, because most folks are not thinking about preventing death, but about maximizing life. It would certainly seem odd if one went up to the parent of the helmetless child and declared, "The most likely cause of death in children is accidents! Get a bike helmet!"

As physicians, we refer to specific patients, families, or whole populations as being at "increased risk for . . ."—usually a particular illness or disease. Some of these risk factors are based either on environmental exposures, such as polluted cities, living close to a major roadway, cigarette-smoke exposure, specific factory work, or working in a health-care environment. Cigarette smoking has traditionally been the easiest risk to quantify. We count cigarettes in pack years. A twenty-year history of smoking two packs per day is a forty-pack-year smoking history (20 years × 2 packs/day = 40 pack years). The higher the pack-years number, the higher the smoking-related risks of cardiovascular disease, cancer, and chronic lung disease.

Alcohol consumption is a bit less easy to calculate, as there are so many types of alcohol (wine, beer, liquor) and consumption tends to be more variable. From a medical standpoint, we start entering the alcohol-related risk factors when people describe their consumption in bottles versus glasses of wine, cases versus cans/bottles of beer, and fifths versus individual drinks of hard liquor. We don't split hairs over two or four glasses of wine per week. But when we ask how much beer is drunk and the question returned is "You mean cases per week?," we take note and factor that into risks of liver disease, liver cancer, car crashes, and even domestic violence. Risks for particular illnesses or diseases also vary not just on consumption of cigarettes, alcohol, or drugs, but also on where one lives, the family situation (married or single, for example), or type of living quarters. Environmental risks can also be related to use of drugs (including over-the-counter, prescription, and illicit) and sun exposure.

Exhaustive research has clearly defined the health risks of cigarette smoking, alcohol abuse, drug abuse, and sun exposure. Few doubt that cigarette smoking puts you at increased risk of developing lung cancer, throat cancer, heart disease, peripheral vascular disease, and chronic lung disease. Few would disagree that alcohol abuse causes liver disease, liver cancers, throat cancers, dementia, and vehicular deaths. Many may not know that the combination of cigarette smoking and alcohol abuse increases many of these risks multiple times over, not just additionally. Most are quite aware that misuse of prescription drugs can lead to devastating outcomes. This is also the case for over-the-counter drugs. An overdose of Tylenol, the brand name for acetaminophen, can kill you. Even I, a self-proclaimed sun worshipper, know that sun exposure significantly increases risks of developing all types of skin cancer, including basal cell, squamous cell, and melanoma. These statements are not hype. There's hardly controversy. They are barely news.

Many hyped health risks don't actually exist, such as drinking coffee while pregnant, giving multiple immunizations at once to an infant, and drinking tap water in an industrialized city. But perhaps more concerning are the seemingly less risky behaviors that are becoming more worrisome and risky than the well-described sins that too many of us still have. In 2014, Oxford Dictionaries stated that *vape* was their Word of the Year. A close runner-up was *budtender* (someone who serves marijuana at a pot bar). To give you some perspective, ten years prior the Word of the Year was *defriend*, and ten years prior to that it was *fashionista*. Interestingly, a hundred years prior to *vape*, the Word of the Year was *environmentalism*. My, have we come full circle—from cleaning to polluting in one century. To vape is to inhale and exhale the vapor from an electronic cigarette.[12] What *vape* and *budtender* have in common is that they both relate to smoking, and they are both suddenly again trendy, hip, and erroneously presented as safe, as were cigarettes in the 1940s and '50s.

Electronic cigarettes, or e-cigarettes, were first explored in the 1960s.[13] The goal was to create a cigarette whereby one could inhale nicotine without inhaling tobacco smoke. Let's pause for a moment and discuss why this is just not okay. Indeed, nicotine is less carcinogenic than numerous chemicals in traditionally inhaled and chewed tobacco products. But nicotine is a highly addictive drug. Inhaling nicotine from an e-cigarette causes the same effect, including addiction, as inhaling nicotine from a traditional cigarette. While e-cigarettes were first deemed and marketed as safe cigarettes, this is far from the truth. Indeed, they have been found to improve the likelihood that those cigarette smokers who had tried quitting in the past will quit. This is significant, as cigarette smoking is responsible for nearly 6 million premature deaths worldwide annually.[14]

But e-cigarettes are not the candy-colored, fruity-flavored kiddie treats that one might envision. While great strides are being made in getting people to stop smoking by using e-cigarettes, they are simultaneously creating mini-Marlboro girls and boys. Not only do they carry inherent risks of nicotine exposure and addiction, but early studies have demonstrated increased use of traditional cigarettes by young people who began by vaping e-cigarettes. The initial marketing push was monumental, with ads and designs clearly aimed toward the middle school set. As with many large-scale novelty items, the marketing worked. Annual National Youth Tobacco Surveys of children in grades six to twelve were compared from 2011 to 2013. Over this three-year period, the number of first-time users of e-cigarettes increased threefold, from 79,000 to over 263,000. When asked if they intended to smoke traditional cigarettes, 49.3 percent of users of e-cigarettes said yes, compared to 21 percent of those who had not vaped. Even for those who had tried e-cigarettes just once, their odds of going on to smoking tobacco cigarettes almost doubled compared to those who had never tried them. Electronic cigarettes have become the learner's permits for smoking.[15]

Speaking of learner's permits, driving a car has always been considered to be fraught with risk—driving while intoxicated is probably the most widely known risk, and concern, for young drivers, increasing their chances of injury and death astronomically. The new drunk is texting while driving (TWD). Not only does this increase risk of crashes, but TWD has been linked to other risky driving behaviors. That's right. We are now fully aware that texting while driving, by either adults who've been driving for years, or by new drivers, increases risks for disaster. But what many may not know is that a prior history of texting while driving increases likelihood of driving while intoxicated, being a passenger with a driver who is intoxicated at the wheel, and erratic use of seat belts.[16] So not only is texting in and of itself a problem, but it is linked to other risky behaviors. This is similar to the pattern for e-cigarette use and traditional cigarette use. One risky behavior leads to another. Texting is the proverbial gateway drug to drunk driving.

Marijuana is no longer the illicit recreational drug that it used to be. It has entered backyards (legally), pharmacies, hospitals, and hospices. Since 1996, the list of states legalizing the medical, and even the recreational, use of cannabis-containing products has been growing. You can now chew it, drink it, and, yes, still smoke it to help relieve pain, nausea, headaches, or even seizures. Countless patients with acute and chronic illnesses have reaped the benefits of this herb. It is safer and less addicting than narcotics, has fewer side effects than many powerful medications, and is rarely abused by patients who need it. Even for recreational use, it is safer than alcohol.[17] But just as e-cigarettes are the learner's permits for tobacco cigarettes, medical marijuana may be the learner's permit for marijuana abuse. Yes, it's better than alcohol, but it remains the lesser of two veritable evils.[18]

While marijuana and alcohol may be fading in the spotlight of drug killers, the opioid epidemic is an increasing death sentence for all ages. Drug overdoses now kill more annually than did HIV/AIDS during peak epidemic years, and the numbers continue to rise.[19] Such overdoses kill more annually than breast cancer, car accidents, or gun deaths and gun homicides. Doctors are part of the problem, increasing their narcotic prescriptions each year for patients with both acute and chronic pain. While the numbers differ from state to state, the United States is, by far, the world's highest consumer of opioids. And opioid abusers do not fit the drug-abuser stereotype that you might conjure. These are folks with visceral pain, musculoskeletal pain, postsurgical pain, and cancer pain. They all need pain control, and resorting to the aspirins and ibuprofens of the world just won't cut it. We prescribe narcotics for children and babies. I know this sounds absurd, but their pain is just as real as an adult's. However, those who treat children have become more and more reluctant to prescribe narcotic pain medications, as there have been cases of sudden death, respiratory complications, and even sharing meds (meaning a parent may swig a few doses of liquid oxycodone meant for their child).

When I was a surgical intern, we would have contests to see who could fall asleep fastest while standing up. We would fight to get to be in an operating room where we could "water-ski." To water-ski we would hold a surgical retractor with just the right combination of tension and stability to hold us up like a water-ski handle. This was the best, and least visible, way to get a good snooze during a case. After my own solid snooze while waterskiing, I would call my chief residents with a question in the middle of the night. They'd answer quite lucidly on the phone, but have no recollection of the call the next morning. They'd answered in their sleep. Sleep deprivation was a rite of passage, a badge of honor, a shared code, of residency. Sleeping was for wimps. For the regular humans. Certainly not for the surgeons. A young college student named Libby Zion changed the view of sleep deprivation forever.

In March 1984, Libby went to a New York City hospital emergency room for what appeared to be a flu-like illness and muscle aches. She was treated by an intern and a first-year resident and was given a medication to relax her. This drug, Demerol (or meperidine), is one of the most commonly prescribed drugs in hospitals for pain control. Unbeknownst to the sleep-deprived residents, Libby had been taking a prescription antidepressant called Nardil (phenelzine), a type of antidepressant known as an MAO inhibitor (monoamine oxidase inhibitor). Patients taking MAO inhibitors usually have not responded to more standard antidepressants. MAO inhibitors carry known risks, and users cannot eat certain cheeses, dried fruits, or cured meats, as these can interact with the drugs and cause life-threatening high blood pressure. MAO inhibitors can also interact with opioids such as Demerol, leading to confusion, muscle contractions, fever, and death. The two drugs used simultaneously in Libby Zion resulted in a fatal reaction, and she died that night of a cardiac arrest. Had the treating doctors been aware of Libby's use of Nardil, they would not have prescribed Demerol, knowing the risks. Besides bringing civil charges against the physicians and the hospital, Libby's father, an attorney and journalist, raised the issue of inadequate sleep and insufficient supervision for medical trainees. While it is common practice to ask a patient what medications they are taking, Libby was not asked, or perhaps she did not offer this information. The use of psychiatric medications, even nowadays, remains somewhat stigmatized. It is unclear whether Libby felt too sick to share this information, whether the residents were too focused on her treatment to ask, or whether she was asked and denied medication use. Even today, many who use herbal supplements and homeopathic nonprescription medications do not inform their doctor of this as they do not consider these to be medications at all. Because of Libby's case, in 1989 resident work hours were reduced to eighty hours per week maximum, and no more than twenty-four consecutive hours. This did not apply to surgical resident hours, although the Accreditation Council for Graduate Medical Education has gradually instilled work-hour restrictions in all specialties. A program that violates these restrictions can lose accreditation.

Although we who trained in the days prior to this nap-and-hot-cocoa time, as we like to call it, and rib the current residents on how easy they have it, it's no joke. Did sleep deprivation kill Libby Zion? We may never know. But years later, the perils of sleep deprivation became apparent, and sleeplessness was no longer cool. It was dangerous, unhealthy, and foolish. It is now well-known that short-term sleep deprivation is associated with poor performance, dangerous driving, increased stress, and overall irritability. Prolonged sleep deprivation has far more negative consequences, from memory loss to hypertension and higher risk of certain types of cancer (see below). Even today, you can ask any intern or resident about his or her lack of sleep—despite strict work-hour restrictions and sleep requirements, based primarily on the negative medical decision making caused by inadequate sleep—most interns and residents remain sleep deprived. I know this is true, as they all somehow sleep through my dynamic, exciting, edge-of-your-seat-suspenseful lectures. It *is* because they are sleep deprived, right? Professional truck drivers need to get adequate sleep to do their job safely, as do pilots, train conductors, and, of course, surgeons.

But aside from the obvious risks to others that come with being tired or overtired, are there medical risks to unrelenting sleep deprivation? Absolutely. Sleep deprivation has been found to cause increases in blood pressure and decreases in glucose metabolism, leading to insulin resistance and increased risk of diabetes. Leptin, a hormone that signals satiety in the brain, is decreased with sleep deprivation. Ghrelin, a hormone signaling hunger to the brain, is increased with sleep deprivation. The combination of insulin resistance and hormonal imbalance leading to hunger may lead to sleep-deprivation-associated weight gain. It's not just the midnight snack and the comfort food after a late night that do it. There is also an increased level of inflammatory mediators in those who are chronically sleep deprived. The combination of higher blood pressure, increased appetite, poor insulin metabolism, and chronic inflammation all lead to increased cardiovascular risks.[21]

The environmental risks on which I focus are not pollution, living situation, or toxic cleaning chemicals, but those risks to your body from your own microcosmic environment. You can control these factors, which don't require policy change, label reading, or moving out of your apartment. We are so focused on our environment globally that we easily forget the environment surrounding our own selves.

THE POWER—OR NOT—OF GENES

Genetic risks are equally as important as environmental and behavioral risks, and no less multifactorial and complicated. For anyone who's gotten beyond middle school genetics class, we know that it's not just as simple as brown eyes, blond hair, and height. Genes are not as simple as those Punnett squares with Bb and bb coming out to 50 percent chance of dominant and 50 percent chance of recessive traits. Genes can be hidden for generations or be more powerful than we would ever want. What does it mean to have good genes, and can they be modified? And if so, would that make us genetically modified organisms (GMOs)? If we can't change our genetic makeup, do we want to know what we're at risk for?

Let's start with a genetic factor that is common, yet certainly not dangerous or harmful to our health. Beginning in 1997, all babies born in the United States were required to have a hearing screening before leaving the hospital. Those who failed the screening were referred for further testing by an audiologist. Universal hearing screening has been life-changing for countless children—those whose hearing loss (unilateral or bilateral) would not have been discovered until kindergarten or later was found in the first three months of life. If children with hearing loss receive hearing aids in those first months, there is substantial reduction in speech delay, learning difficulties, and social isolation. Widespread screening has led to the discovery that approximately one in three hundred healthy babies born in this country has some hearing loss.[21] While it is not headline news, hearing loss is the *most common* congenital anomaly. And the most prevalent reason for this variation is a gene. This gene is not the typical brown hair/brown eyes gene, but a weakly expressed gene for proteins that may go hidden for generation after generation before it appears in a child with hearing loss. These proteins, called connexin 26 and connexin 30, can be identified by genetic testing.

Otherwise healthy babies (meaning babies with no other medical issues except for hearing loss) are often tested for this gene. Sounds simple, yes? Not so fast. Hearing loss has been considered a deficit in some eyes, and a trait in the eyes of others. In the years surrounding the advent of cochlear implantation, a surgical procedure followed by intense therapy to provide some degree of hearing to those with none, the deaf community was in an uproar. Sign language is a language, after all, and the idea that those who use it are inferior is controversial. Even putting this controversy aside, let's say we test a baby with hearing loss for the gene, and we find it. The good news is that we have an answer—a reason—for the hearing loss. But do we test the parents? Do we test the siblings, who may have normal hearing but carry the gene and may pass it on? And do we tell the baby when he or she gets older? Let's say the baby has mild hearing loss in one ear, but doesn't require hearing aids. Are his parents required to tell him of this genetic finding when he's older? Does he tell his future partner? And is hearing loss a true deficit, or a trait variation that many embrace? The gene pool is deep. Swim carefully.

The ethics of genetic testing become even more complex with a serious disease. Hard choices will increasingly be made in modern medicine now that we have technologies to identify genetic forces. For example, consider the following real-life situation: The husband of one of my colleagues has Huntington's disease, a devastating illness that robs people of their bodies and minds—rendering people unable to walk, talk, or even feed and take care of themselves. It usually doesn't develop until middle or late adulthood, usually with slowly progressive symptoms early on, followed by a roller coaster of progression. Medicines can temper some of the symptoms, but there is no cure. It's caused by an inherited genetic mutation that leads to degeneration of nerve cells in certain parts of the brain. Some describe it as Alzheimer's and Parkinson's mixed together. As most people aren't diagnosed until they experience symptoms in the prime of their life, many have already had children and potentially passed on the defective gene.

My colleague has three biological children, ages ten, twelve, and fifteen. When do they get tested? Each child carries a 50 percent chance of having the disease. Because the gene is dominant, a person only needs one copy of it, from either parent, to have the disease. The three children are witnessing its effects on their father (and his mother) as they grow up. Every time one of the kids trips on a step, misses a pass at a football game, or forgets a homework assignment, the parents wonder if Huntington's is around the corner. Now that the children are getting older, they are wondering, too.

IN-HOME GENETIC TESTING

The past decade has brought the advent of in-home genetic testing kits. You can even purchase two brands of such tests, as well as one for your dog, for under $300 on Amazon.[22] Is the now-basement price of $99 per kit worth even that little? Let's put your pooch aside for now and go through how these work, and what they can show. Before we even consider the impact of the results on your health, think about the release of personal data when using these kits. Just as your internet search interests can be accessed, leading to your future searches being riddled with focused ads, these companies are also provided with an open window to invade your privacy.

One of the early companies, established in 2007, was 23andMe. You spit in a vial, send it in, and they will tell you such fun facts as the nature and quality of your earwax, whether or not your pee smells funny after eating asparagus, and whether looking into bright light makes you sneeze. I don't know about you, but I know these highly personal details about myself for free. The upside is that this company and others are using compiled personal data to help find genes associated with disorders such as Alzheimer's. While your personal data may not seem to carry much significance to you, the longer-term benefits of obtaining large amounts of our population's genomic code may have substantial impact in researchers' ability to identify, treat, or even prevent certain genetic disorders. But 23andMe will also tag individual results and send you ads based on its genetic findings about you and perhaps share these findings with health insurers and pharmaceutical companies. The company states on its site, "Genetic Information that you share with others could be used against your interests. You should be careful about sharing your Genetic Information with others." And you thought Google was chasing your searches uncontrollably.[23]

Is there any benefit to in-home, or direct-to-consumer, genetic testing? Well, if the mild variant of hearing loss can be met with controversy, whereby we require extensive counseling with a trained genetics counselor for a baby's parents before testing to ensure they understand its implications, you can be assured that the homemade spit vials are certainly suspect. But the goal of in-home genetic testing is enticing—who wouldn't want to know his or her genetic risks of heart disease, stroke, certain cancers, and so much more, with the idea that somehow changing one's lifestyle, seeking earlier screening, and even beginning treatment and prevention early can stave off these indolent maladies?[24]

Here's how it works. The Human Genome Project, completed ahead of schedule in 2003, identified the 3 billion possible permutations that make up human DNA, the building block for our genes. Interestingly, we diverse humans share 99 percent genetic similarity with one another. While that sounds high, it leaves over 10 million variables open for differing genetic makeups. Genetic testing for known factors such as the breast cancer gene (BRCA1) or the cystic fibrosis gene is not so simple that one gets a yes or a no on a laboratory report. Carrying BRCA1 has many implications, including that having the gene does not necessarily have any bearing on future disease. So many people receive this test not knowing that it is simply a marker for a potential type of breast cancer (or even pancreatic cancer), but that it does not necessarily mean that you are destined for cancer. The same goes for so many genetic test results—some genes are stronger than others, meaning that having a gene may not have any bearing on the actual trait, or illness. Genes carry with them wide variables, varied disease manifestations, and differing monitoring and treatment recommendations, based on more than just a positive or negative result. Clearly delineated results from these companies can be misleading at best, and inaccurate at worst. Results may lead you to either unnecessary or inadequate diagnostic testing, adding unnecessary costs and sometimes high-risk procedures that could easily have been avoided. Save the questions about your family's health history and your own risk profile for a health-care professional whom you trust. Genetic testing should ideally include genetic counseling. If you want to know if your pee smells after eating asparagus, then go for one of those testing kits. But I'd imagine you already know that without having to submit your private data, email address, and credit card number to an outside vendor. Counseling not included.

Having a little information, even if it *seems* like a lot of information, such as your family's complete genetic makeup, can have little impact on day-to-day health choices. Yes, genetics play a role in lifetime health risks, diseases, and even traits such as athleticism, good looks, and smarts. But genes take a backseat when so much else is at stake—health choices, lifestyle choices, and even hard work. When Angelina Jolie publicly announced her choice to undergo a bilateral mastectomy because she was BRCA1 positive, the world learned what BRCA1 was, as if it were the latest acronym to hit the medical world. It had been discovered decades before Angelina's decision. The knowledge of a risk does not make that risk bigger. Ebola virus did not hit the planet in 2014. It had been around for decades. Its epidemic did not make it more virulent than it had always been; it just made us more aware of an entity that already existed. Cars, on the other hand, are prevalent, mundane, but deadly. Over thirty thousand people die each year in car accidents in the United States alone. The largest Ebola virus outbreak, which occurred over a two-year period, resulted in approximately eleven thousand deaths worldwide (and, as mentioned, that included only one person in the United States). Approximately forty thousand women die each year from breast cancer in the United States alone, regardless of carrying the BRCA1 or BRCA2 mutation (surprise: the vast majority of women diagnosed with breast cancer have no family history of it).[25] While the mutation increases one's lifetime risk of breast cancer, carrying the gene mutation is not a breast cancer diagnosis. All of the data in the studies, and all of the news one hears, do not replace an individual situation, including risks of elective surgeries, and even risks of diagnostic tests.

Indeed, genes are powerful. They make up who we are—what we look like, our predisposition to disease, our musical ability, athletic clumsiness, or innate smarts. But we are not powered by genetics alone. Twin studies, looking at identical twins separated at birth, have demonstrated both the power of genetics (nature) as well as the power of one's environment (nurture). Many of us have known rising stars, in elementary school, high school, or beyond, who, after some untoward turn of events, went sour. We also know some success stories that developed simply by virtue of hard work, not natural talent. All of us carry powerful traits, many of which we try to utilize or sequester. Most often, however, our environment molds our given traits. You may have no genetic predisposition to cardiovascular disease, but if you smoke, don't exercise, are obese, and live a stressful life, your good genes go by the wayside. You may have a strong family history of high cholesterol and heart disease, but if you manage your diet, take recommended medications, and exercise, your cholesterol can be just as normal as that of one who has no risk for high cholesterol.

So here's the answer to the question I posed at the start of the chapter, What do Ebola and your car have in common? Both can kill you. But one is far more likely to be a factor in your end than the other. Even if you were to travel to an Ebola-outbreak area today, your chances of contracting the virus are still far less than being killed in a car accident in America this year. Buckle up, and breathe a sigh of relief.

HYPE ALERT

* Worry more about getting into a common car accident than uncommon germs such as Ebola.

* When people worry about the risks of a certain procedure, treatment, or surgery, they often fail to consider the potential risks of *not* taking any action—the latter of which could be the riskier choice.

* Every decade of life bears its own set of risks. For about the first half of your life, accidents are more likely to take you out than anything else. Then, as you get older, heart disease and cancer become the more likely killers.

* Your genes speak toward your health risks, but your environment and the choices you make (e.g., to smoke or not to smoke) speak louder.

* In-home genetic testing is not what it's cracked up to be (yet). Proceed with caution.

3

TURF WARS: AN IMPORTANT LESSON IN CORRELATIONS

How to Understand *Cause, Link,* and *Association*

HOW MUCH SHOULD YOU WORRY ABOUT CHEMICAL EXPOSURES?

DOES *LINKED TO* MEAN THE SAME THING AS *CAUSES*?

DO CELL PHONES AND ARTIFICIAL TURF CAUSE CANCER?

WHAT'S THE LINK BETWEEN SNORING AND ADHD?

HOW BAD IS PLASTIC?

When shopping for beauty products and toiletries, you're more likely to find labels such as *paraben-free* and *phthalate-free* today than ever before. Why? What's the big deal? Are these ingredients harmful or hyped?

The terms *linked to* and *associated with* can be deceptive in the media when they are taken out of context. These terms are, by definition, speculative. In medicine, we call this *true, true, and unrelated.* We see two events in the same setting, but one does not necessarily cause the other. The only link may be that they occur at the same time and same place. The relationship may be purely coincidence. Another way to think of this is to consider the following example, albeit absurd, but it drives home the point: In the past decade, gym memberships have increased and the number of people running marathons has surged.

But there has also been a sharp rise in obesity. Do gyms and marathons cause obesity? I think not. The term *correlation* describes the strength and direction of two events, and a correlation may be positive or negative. An example of a correlation would be that good weather has a positive correlation with the number of beachgoers, whereas bad weather has a negative correlation with the number of beachgoers. *Link* has the same meaning as *correlation:* committing a crime is linked to going to jail. It's not definite, but one event is often followed by the other. Committing a crime does not *cause* you to go to jail, but the two are clearly linked. Causation is the most challenging to prove. This requires 100 percent correlation. Although we are comfortable stating that cigarettes cause lung cancer, even this is not true in the pure statistical sense of the word *cause*. For this to be true, 100 percent of cigarette smokers would have to develop lung cancer. But even without 100 percent correlation, causation in the clinical sense can absolutely be said, especially when it comes to smoking.

There is little doubt these days that cigarettes cause all types of cancer. But in the 1960s this was still unproven. It was already recognized, from several studies, that cigarette smoking was associated with an increased incidence of lung cancer, but it was still not referred to as the cause of the cancer. I remember my grandfather, around that time, asking me to pass him his "coffin nails." A 1965 presidential address to the Royal Society of Medicine could be applied to today's world in contemplating when it's okay to jump the gun on causation, and when things should be taken in stride as a likely coincidence.

Sir Austin Bradford Hill was an English epidemiologist and statistician whose work throughout the twentieth century pioneered the gold standard in testing theories: the randomized clinical trial. His work helped establish the connection between cigarette smoking and lung cancer. In his 1965 address at the Royal Society of Medicine, he raised the question of association versus causation when he wisely surmised, "Before deducing 'causation' and taking action we shall not invariably have to sit around awaiting the results of [the] research. The whole chain may have to be unraveled or a few links may suffice. It will depend upon circumstances."[1]

In other words, when issues such as time frame, known biological factors, consistency of results or findings, and specificity of an already-recognized association come into the picture, the issue of causation over association can move up the chain more quickly.

Hill looked at other areas besides cigarette smoking, including factory chemicals no longer used today because of their eventual clear association (or link or cause) with cancer rates, and even stress as a cause of illness. Hill describes a perfect example of jumping the gun on cause versus association when he discusses a group of peptic ulcer patients treated in the late 1950s. On admission to the hospital, a survey about stress at home prior to admission was given to urgently seen peptic ulcer patients. I know, who would want to take a survey when they are suffering from an acute peptic ulcer? In the 1950s in the United Kingdom, this was the norm, and the patients did as they were asked. Stiff upper lip. Their results were compared to the same survey given to non-emergency hernia patients coming in for elective surgery. It's no surprise that the ulcer patients reported more stress than the hernia patients in the prior weeks. But regardless of their stress, they were compelled to come to the hospital to address their aching ulcer symptoms. The patients with hernias that didn't bother them and who may have had an equally stressful week could safely stay at home and recover from their stress before getting treated. They'd go to the doctor when they were no longer stressed by outside issues. Which invites us to question, Did stress cause the ulcer or was it just bad timing?

SUPERSTITIONS, FUZZY SCIENCE, AND MISNOMERS

In some medical and surgical circles, even *we* can be embarrassingly superstitious about cause and association, blaming (gulp) supernatural forces for strings of untoward events: "I never operate if there's a full moon." "Never perform an elective procedure on Friday the thirteenth." "Redheads have more bleeding. Surgeons beware." "Complications come in threes." While this is all a load of hooey, it had enough traction to lead one academic group at Temple University in 2004 to go ahead and definitively prove these unfounded theories wrong when it comes to taking children's tonsils out.[2]

I remove tonsils regularly in my work, but I do have my own superstitions. I like to think some of this is just habit, but a little supernatural thought is thrown in, for sure. For instance, I always remove the right tonsil first. Always. Not that it's technically different or challenging to remove the left one first, but the right is always first. As I tell my OR team, "I'm always right." I never tell the patient's family, "Everything will be fine." I feel saying that could be bad luck, as I'm saying something I cannot possibly know will be 100 percent true. Instead, I say, "We'll take good care of her." This is 100 percent true, and I feel less worried that way. I never dictate my operative report until the patient is awake. While this is also just plain old good medical practice, I feel that if I say something that hasn't yet happened, such as "The patient was awoken and taken to the recovery room," something bad might happen. And then there are the socks. Yes, the socks. Surgeons wear clogs in the operating room—they are comfortable for standing for long periods. As our scrubs are drab, and clogs are simply practical, I've taken to wearing fun cycling socks in the OR.

In the early 2000s, a critically ill newborn came to the operating room with life-threatening airway anomalies and breathing problems. I had the A-team in the OR that day: my favorite anesthesiologists, superb senior residents, and the best nursing team. The procedure was going as smoothly as we could hope for, until everything turned south. After the tracheotomy was placed, the baby's lungs shut down. She could not be ventilated. We tried to resuscitate her for what felt like seconds, but turned out to be several hours. In came pediatric surgeons to drain the lungs and place central and arterial catheters, as well as cardiac anesthesiologists to monitor the heart. No fewer than twenty specialists surrounded and treated this baby, until I did what every doctor dreads: I called the code. As I announced the time of death, with date, hour, minutes, and seconds, the room went silent. She had died. That day, I had worn my favorite cycling socks, with goldfish on them. I still have those socks, perhaps out of respect for this child's six-day life. But I have never worn those socks again, and I never will. Superstitions can be based on the fear of the supernatural, but mine and so many others are based on the fallacious notion that our own acts somehow carry more power than they do.

Thankfully, the authors of the Temple study found no increase in bleeding after a tonsillectomy based on the phase of the moon, the date on the calendar, or the color of the patient's hair. And triads of complications did not occur. We even carry these superstitions to more life-threatening illnesses, outside the surgical arena: "Bad prognostic sign: the family is nice." And probably one of the least politically correct phrases bantered around the physician circle: "You can't kill sh*t" (meaning that nasty people seem to live forever, but the nice ones die).

Few physicians, scientists, or academicians will come out and say, "X causes y," unless reproducible large-scale studies document a connection. It's safe to say now that tobacco can cause lung cancer, but we can't say that aluminum-based deodorants cause breast cancer (at least not yet) or that coffee stunts your growth when consumed during development. We can say that excessive sun exposure and tanning beds can cause various types of skin cancers, but we can't say that hair-straightening products cause scalp cancer.

A "cause" may seem to take place in a laboratory setting, but not be found in humans. One can easily find ways to demonstrate that ingredients are toxic in a laboratory setting, but that doesn't necessarily translate to the human body. A remarkably simple example of this is good old table salt, with the chemical name sodium chloride, NaCl. Putting aside the heated debates over the pros and cons of salt in one's diet, its two simple ions—sodium and chlorine—are, as stand-alones, two of the most deadly elements. Sodium is a shiny, buttery substance that, when it comes in contact with water, explodes. It doesn't just kill cells; it explodes laboratories. Chlorine is the substance in swimming-pool sanitizers, and bleach. Lots of other ingredients can be toxic in a laboratory setting at a high enough dose. Some of these ingredients are thought of as "healthy" at the right dose: fluoride, vitamins, omega-3 fish oil, and alcohol. Apples, wine, water, and rice contain minute amounts

of arsenic, which is toxic in large amounts. Pears contain formaldehyde, magnitudes more than any vaccine. The same is true in reverse:
just because a substance found in okra has been found to annihilate
human breast cancer cells in vitro (in a glass petri dish) doesn't mean
okra is now a proven anticancer medicine.

Fears surrounding parabens and phthalates in cosmetics and lotions, for instance, gained traction from studies done on cells grown
in tissue cultures outside the human body (in vitro). But that's not a
fair equivalent to a complex human body. To compare the impact of
these ingredients on a single cell type to their impact on the intricacies
of the human system is fuzzy science. Unfortunately, we don't have a lot
of human—or even animal—studies to show us what these chemicals
could be doing to us (in vivo). Does this mean it's wise to be cautious?

While long lists of unrecognizable chemicals are reason enough to be
cautious about putting something into our bodies, we need to remember that chemicals are absorbed differently if rubbed on the skin, inhaled, or ingested. Indeed, the skin does absorb small portions of certain
chemicals in skin-care, hair-care, and cosmetic products, although
even those substances absorbed usually go only skin deep. Rarely, in
such aids as estrogen patches, nicotine patches, and even methylphe-
nidate (similar to Ritalin) patches, medications applied to the skin are
meant to be absorbed into the bloodstream. But skin creams, despite
their long lists of chemicals, barely scratch the surface. Inhaled medications, such as asthma medications, do get absorbed not only into the
respiratory system, but also into the bloodstream. These include medications to open up the bronchi (bronchodilators) as well as those to
reduce inflammation in the airways (inhaled steroids). Bronchodilators, by nature of their action on the vascular system, can increase
heart rate and even cause some jitteriness, similar to caffeine's effects.
Inhaled steroids, when used regularly over many years, can be associated
with (although not necessarily *cause*) slight reduction in growth, leading to up to half-inch reduction in the height reached in adulthood.

One chemical additive, primarily in cleaning products and cosmetics, that has gotten quite a bad rap is sodium lauryl sulfate, or
SLS. This emulsifying agent has gotten emulsified itself by the lay press,

which claims it has cancer-causing properties and can damage organs, the eyes, the skin, and the environment. Large studies have demonstrated no such effects, especially when it's used in such infinitesimal amounts as in consumer products. But the combination of its somewhat "toxic" sounding name and misinterpretation of the data have put SLS on the top of the list of harmful additives in commonly used household products.[3]

So many buzzwords are used in casual speech, all sounding somewhat definitive and scientific, that we take their meaning as fact. *Toxin* is a popular one—we've heard about how toxic our world has become, from processed foods filled with synthetic ingredients to environmental chemicals we're exposed to through the air or products we buy. The words *chemical, additive,* and even *plastic* have such negative (hyped) connotations. On the flip side, words such as *natural fragrance, unscented, organic,* and *natural* are often not as they seem. Use of the terms *organic* and *natural,* for instance, mean different things to different people (and different companies). The FDA does not regulate these terms, so people are free to use them as they wish. To assume *natural* and *organic* equal "safe" is unfortunate; natural ingredients can be toxic, too. Poison ivy is natural, but you definitely don't want to rub it on your skin. Ingredients in beauty products, cleaning products, artificial materials, foods, and drinks are perfect examples of how substances can be muddied by such terms as *correlation, association, linked to,* and *caused by.* When presented in news stories, promotional materials, or warning labels, any of these terms can be quite compelling.

A few years back, I received an urgent call from my child's preschool. The school was planning on removing the artificial turf from the play area and wanted to know my thoughts on the matter. They had heard it caused cancer and wanted it out. I grew up on the East Coast, where grass was easy to come by and droughts didn't exist, and as I didn't play or watch much football, I was never a big fan nor connoisseur of artificial turf, except for having heard about it as the latest and greatest at the new enclosed Houston Astrodome. But apparently it was becoming more and more the norm, not only on football fields, but also on soccer fields, playgrounds, and baseball diamonds. It would save millions of gallons of water annually, would be pesticide-free, and required little or no maintenance.

But one day artificial turf made a different kind of headline. In early 2016, a surge of lymphoma cases among teenagers and young adults who had played soccer hit investigative journalism and the media. Most had been high school or college goalies, who of all the players tend to spend the most time diving for balls on the ground and tumbling in the artificial turf. They had had the most direct contact with the turf. A relatively new type of turf had been the rage, containing a material called crumb rubber. Crumb rubber consists of repurposed tires, shredded to softness and embedded in the artificial grass. The intention was to create a softer surface than the older version of artificial turf had, making the newer recipe easier on the feet and joints as well as the delicate face upon smashing it into the turf during a goal save.

Over eleven thousand playing fields in this country now use crumb rubber. Crumb rubber is just how you'd picture it—black rubber crumbs. Soccer players would routinely come home and shake out the crumbs from their hair, their socks, and their ears. They would dig it out of their fingernails and scrape away the black residue embedded under their skin. There has been some serious concern about potential carcinogens in crumb rubber, so when an unprecedented number of lymphoma cases—cancer of the lymphatic system—were documented in this soccer population, heads turned to the crumb rubber.[4] The rubber does contain some pretty awful-sounding metals, chemicals, and even known carcinogens. But in analysis, the concentrations were so minute that, even when breathed in, the increased risk of cancer after thirty years of direct exposure to these was calculated as one in a million.[5]

Yes, there are carcinogens; yes, there are cancer clusters. And, yes, these substances may indeed increase the likelihood of cancer in those exposed regularly, or those who have an underlying heightened risk for cancer due to genetic factors. But they are not a definitive cause. At least not that we know of. Not yet. There may be a link or, at most, an association. These soccer players were suffering from Hodgkin's lymphoma, which is the most commonly seen cancer in the late-teenage years, with or without a history of turf exposure. Did the turf increase their risks? It certainly is possible. But headlines and clusters are not the same as identifying a direct cause. While these cancer clusters are worrisome, the findings on crumb rubber should lead us to further studies of a potential cause, not a mere correlation. When two entities are relatively common, such as exposure to artificial turf and the most common malignancy seen in teenagers, it is quite difficult to prove cause. Even young kids have become aware of the potential carcinogenic effects of crumb rubber. Rambunctious middle school boys have been known to grab handfuls of the black dust, throw it to a friend, and say, "Here, catch some cancer." My hope is that future research reveals the true risks of ingredients in artificial turf and to use caution in the meanwhile.

CELL PHONES AND CANCER

The average adult can spend more than ten hours per day looking at screens—including those on phones, laptops, TVs, tablets, etc.[6] Overuse and obsessive checking of cell phones (and their social media accounts) will likely soon be a veritable *DSM* diagnosis. The *Diagnostic and Statistical Manual of Mental Disorders* is, as its title suggests, the bible for the American Psychiatric Association. It provides criteria for all presently known psychiatric disorders, enabling mental health professionals, pharmaceutical companies, insurance companies, and even legal organizations to give diagnoses and treatments to patients in a standardized fashion. The first edition was published in 1952, and as the field of psychiatry continues to evolve, updates on definitions and criteria have been published every five to ten years (the current *DSM* is the fifth edition). Recent years have seen a push to label those who are addicted to their cell phones (and are thus called nomophobes—fearful of being out of contact with their cell phone) as having a true mental disorder. Addicted or not, most of us spend a lot of time on our mobile devices, which could have other health consequences.

Many use Bluetooth, speakers, or earbuds when using a cell phone. Most do this either out of convenience, as a safety feature while driving, or to keep their hands free during multitasking. But some use hands-free to prevent brain cancer. A multitude of studies, in neurosurgical journals, bioengineering journals, and cancer journals, have tried to assess long-term cell phone use and its possible correlation with, link to, or cause of brain cancer. One report went further by showing a connection between cellular radiation and heart cancer.[7]

Many neurosurgeons I know will no longer put a cell phone directly to their ear. Even with these academic studies, some even performing studies of studies, there is, as yet, no consensus on a direct link between cell phone use and brain cancer. Some studies assure us of no increased risk of brain cancer with frequent cell phone use. Some report that the risk doubles over ten years. Some say that the risk increases by 5 to 10 percent. With such widely variable results, how can one possibly make decisions for oneself, family members, and the ever-growing population of juvenile phone users? What about phones on wrists? Does this increase bone cancer from direct exposure of low-frequency radiation to the wrist?[8]

The data being discrepant, the cell-phone/brain-cancer connection hit the airwaves like wildfire in 2016. World-renowned neurosurgeons debated on live TV, each with strongly opposing views. Who was right? As with many evolving technologies and variable findings of disease frequency, we still don't know.[9] In the grand scheme of evolving technology, cell phones are still relatively new. Brain cancer is not an isolated event that can be purely related to one type of exposure—cell phones or otherwise. I still use my cell phone, oftentimes directly on my ear. And I may get brain cancer, not from cell phone use, but from some as-yet-unknown genetic predisposition, or even bad luck. Or, yes, maybe (partially) from my cell phone. But the jury is still out. For now, we definitively know that cell phones *cause* more car accidents than brain tumors. Offensive texts and hurtful social media *cause* more brain pain and even suicide than the electromagnetic waves through which they travel.

PLASTICS

Read the list of ingredients of most skin moisturizers, shampoos, and even plastic containers. If you've lived throughout the past several decades, you may have noticed that the creams are getting softer, the shampoos make your hair shine more, and the plastics are more flexible yet more durable. This is all thanks to the many additives being used in these products, including phthalates and bisphenol A (BPA). These new ingredients are responsible for ridding us of glass bottles and helped to create every form of baby-bottle nipple and sippy cup in the first decade of the new millennium. They also aided in revolutionizing neonatal intensive care units by providing more durable, yet flexible, plastic materials for breathing tubes, suction catheters, intravenous tubing, and ventilators. About ten years into their use, concern arose regarding their potential carcinogenicity, as well as their potential action on the endocrine (hormone) system in humans.[10]

These compounds, as well as their metabolites, can be found in human urine, breast milk, and even cord blood. They can be found in semen, and some have proposed that increased presence of phthalates may alter sperm morphology, leading to male infertility or birth defects in children. Some have hypothesized that they increase the risk of type 2 diabetes. BPA levels have also been found to be higher in women with multiple miscarriages than in those without. All of these studies have been useful, but all share some things in common: they are small (all less than one hundred subjects) and oftentimes don't take into account outside variables, such as that people often volunteer for these studies, which can create bias and not reflect an ideal population.

What's more, it can be hard to reproduce the data generated, and other factors may account for these outcomes. We refer to these other issues as confounding variables. For instance, a woman with high BPA levels has multiple miscarriages: the study found that the mean BPA level was higher in women with multiple miscarriages, but the median BPA level was the same in their group as in the group with normal pregnancies. What other factors were taken into account? Maternal age? Paternal age? History of gynecologic illness prior to the study? History of irreg-

ular menses? It's hard to find a cause when other potential contributing factors are not accounted for. This is a common hurdle in human
medical studies: it's difficult to tease out cause versus association. While
most good medical reports acknowledge this, even the words *might
be associated with* or *could be linked to* become headlines.

As we consumers became savvier about potentially harmful additives, we sought confirmation that the substances were removed, somehow convincing ourselves that we would be healthier for their demise
in our products. BPA, for example, is perhaps one of the most notable
toxins.[11] This compound, which mimics some sex hormones, has been
found to be potentially detrimental in both human and rodent studies.
The store aisles with large-print BPA-FREE on plastics are now a must,
both in the baby-supply section and the kitchenware section. But here
is an emblematic example of a false sense of healthfulness.

First, many products that one would not have considered to contain BPA, such as canned goods, water-supply pipes, and incubators for
newborns, in fact do. Second, the plastics manufacturers have managed
to flaunt BPA-free by substituting for it its cousin BPS, or bisphenol S.
The chemical structure of BPS is quite similar to that of BPA, giving it
similar qualities to improve plastics and food preservation. It also likely
has similar effects on the endocrine system as BPA, but has yet to be
demonized to the point scientists are forced to find a gentler substitute.
Studies on BPS are nowhere near as exhaustive as those on BPA, so it
is not considered to be a known hazard. The European Union and the
United States banned the use of BPA in baby bottles in 2011 and 2012
respectively, but BPS remains in the mix. No matter what you're buying,
BPS is embedded in your paper receipts as both BPA and now BPS are
used as developers in many forms of paper, most notably receipts. But
BPS safety is also being called into question.[12]

While the word *plastics* has become equivalent to "toxin," "poison,"
or "archaic additive," these substances date back to 1600 BCE, when
naturally occurring rubber was first found. They are polymers, which
means simply that they are made up of several units (monomers, such
as bisphenol A) in molecular chains. Polymers that are stiffer, made
up of stronger bonds, do not change with heat, but those with more

flexibility, allowing for molding and softer quality, are thermoplastics and can soften when exposed to higher temperatures. In the 1800s, vulcanized rubber and polystyrene were discovered, and the first synthetic polymer, Bakelite, was produced in Europe in the early twentieth century. In today's world, over 300 million tons of plastic is produced annually. It's not going anywhere. And while environmental scientists are acknowledging that such substances as BPA (and probably BPS) can alter endocrine (primarily estrogen) activity, the debate continues regarding the significance of this.[13]

THE SWEET SMELL OF CHEMICALS, OR NOT

I went to sleepaway camp in the 1970s. I can still smell the waft of Herbal Essences shampoo, almost as a visible kelly-green whorl, from the girls' showers around dinnertime. Just conjuring up that image takes me back to a time of innocence and bliss. I also remember that Bonne Bell lip balm not only smelled like raspberry, Coca-Cola, or bubble gum, but that these fragrance-filled chemical sticks we rolled onto our lips tasted like their smells. Every tween girl I knew would snack on her lip balms by licking her additive-laden lips. It was the best. Camp showers now smell of water. And maybe a slight hint of teen sweat. Shampoos are clear, bland in smell. Lip balm is fragrance-free. And definitely tasteless. The flip side of having seemingly dangerous compounds in your day is the illusion of absence of harmful substances. Two cases in point: How many times have you seen the words *unscented* and *fragrance-free* on labels? Some people use such labeled products simply because they don't want their pits to smell like mangoes, nor their hair to smell like an apple orchard. But more commonly, the false notion is that you are somehow being healthier, and more pure, by using substances with such claims.[14]

Products labeled *unscented* may—get this—actually *add* fragrance to block out the scent. And *fragrance-free* may simply mean that no added fragrances were included to block out the already-present fragrance in a given product. Neither term is backed by a legal document or FDA regulation, so their meanings are up for grabs, depending on the manufacturer's angle. This is where we, as consumers, can get a bit duped. We think we are using something healthier when the illusion of healthy is all there is. Phthalates, for instance, have been found in plastics, cosmetics, and lotions since the 1930s.[15] Although they differ in chemical structure from BPA, they are also considered to be endocrine-disrupting compounds, affecting precocious pubertal development of both male and female reproductive organs. Despite some early findings of effects of phthalate levels, both in utero and in early infancy, the debate persists regarding the direct health effects of phthalate exposure. And substances such as phthalates and BPA are just a few of the many plasticizers in products of yesteryear as well as today. They are just some of the few that have, to date, been studied. And labels can be misleading: *unscented, fragrance-free,* or *natural* have nothing to do with presence or absence of substances such as BPA or phthalates. And *BPA-free* and *phthalate-free* have nothing to do with presence or absence of fragrances.

In sum, while data on impacts of all of these additives continues to be found, the main changes have been in wording and in a bit of smoke and mirrors in product labeling and safety. We still don't know for sure how much BPA is okay, how much phthalate crosses the safety border, and what the direct impact on growing endocrine systems will be. Today's safe material may be tomorrow's toxin. We avoided BPA, only to find similar hazards from BPS. Additives are present to prevent other ills. Some additives prevent bacterial contamination; some were created to replace the use of seemingly safe chemicals such as sodium hydroxide (lye, a simple base of cleaning fluids that killed thousands due to its ability to burn human flesh); some were created to simply prevent products from getting moldy in your cabinet.

WHEN CAUSE AND EFFECT ARE KNOWN BUT NOT HEEDED

While I've always been skeptical about what many would consider to be harmful to themselves or the environment (processed foods were made so college kids can afford to eat quickly and cheaply; microwavable plastic is one of the best inventions to date; your floor's not clean unless your eyes tear just a bit), I am scientifically inclined and know what can really do some serious harm. Despite that, even the most type A premed student doesn't always follow her own advice.

I went to college in the tundra we call Ithaca, New York, where we had notoriously cold, snowy winters, but beautiful springs and early summers. Being a sun worshipper who realized, later in life, how much I hate cold weather, those winters were tough on me. But the springs were glorious. You could always find me on a lawn during spring-semester-finals study week. Not under a tree, but basking in the sun, books in hand. I'll never forget studying for my spring-semester genetics final. The book cover was a metallic gray and almost acted as a sun reflector when not in use. The section on the genetic impact of sun damage was fascinating.

The UVB rays of the sun create abnormal bonds in DNA called thymine dimers. These dimers are what cause sunburn, as well as production of the pigment melanin. Some people have better ability to repair these bonds after damage. I somehow convinced myself that I was genetically programmed to avoid skin cancer, due to my strong thymine-dimer recovery. It was a nifty way to remember some genetics while basking away. I went to the final, nose red, even peeling a bit, and proudly told my professor that I was repairing my thymine dimers. She rightly thought I was moronic. It was one of the few B's I received in my premed career.

Sun damage is a well-studied cause of all types of skin cancer—basal cell carcinoma, squamous cell carcinoma, and melanoma. For a time, tanning salons were purporting that they only used the "safe" ultraviolet rays, known as UVA, as opposed to the more notably carcinogenic ones, UVB. It is now clearly recognized that both can contribute to cancer formation, even though the UVB rays are worse.[16] (As the derms like to preach: A is for aging, B is for burning.)

In 2009 the International Agency for Research on Cancer classified tanning beds as a carcinogen. Tanning salons in the Indoor

Tanning Association continue to deny associated cancer risks, claiming damning studies were on those with fair skin only, or claiming that dermatologists are in cahoots with sunscreen companies. Indoor tanning has become more popular in young kids, increasing both short- and long-term risks of skin cancer. Many salons now restrict clients to those over age eighteen. While nowadays I live in one of the sunniest places on earth, even *I* have made more and more effort to turn my back on the sun. Admittedly, I still do enjoy a little more sun than I should—with my SPF 30 and a cool hat and sunglasses. After all, pale is the new tan.

A NEW EFFECT OF AN OLD CAUSE: DOES SNORING CAUSE ADHD?

The past twenty years have seen substantial, remarkable actually, increases in the incidence of both attention-deficit/hyperactivity disorder (ADHD) and autism spectrum disorders (ASD). For those of us who grew up in the 1970s and 1980s, these were the kids referred to either as squirmy or quirky. They got in trouble more than the other kids in the class and were oftentimes labeled class clown or loner. These were innocent, naïve times. Nowadays, the parents of kids who don't make friends easily, or who can't sit still during circle time at preschool (or even prior to preschool), are gently, yet strongly, urged by teachers to get their children evaluated for one of the many so-called alphabet behavioral disorders—those whose initials are now in almost every parent's diagnostic armamentarium, even before his or her child is born—ASD, ADD, ADHD, OCD, ODD, to name just a few. Some of these kids were as young as two. Literally still in diapers. While many of these children did indeed fit the criteria for diagnosis of either ADHD and/or ASD, the question then became—why so many? Were these those same squirmy and quirky kids, or was there a real reason, or cause, to explain this astronomical rise in otherwise healthy kids with psychiatric diagnoses?

In my field, where we treat patients with sleep disorders (usually due to a blockage in the airway), the 1990s brought to the forefront the issue of whether sleep-disordered breathing in children was somehow linked to attentional deficits. Dr. David Gozal was a pioneer in this field. In a small, rudimentary study, he studied groups of six- and seven-year-old children with poor school performance. He did home sleep studies on these children, measuring their oxygen levels and degree of snoring, the latter of which is a sign of airway blockage. He found that a surprisingly high percentage of these struggling kids had sleep-disordered breathing, but that was not the interesting part. The remarkable results came later: half of the kids with breathing problems at night were treated with tonsil/adenoid removal, in efforts to improve their breathing at night. The other half opted out of surgery. One year later, the kids who had had surgery had a statistically significant improvement in their grades and performance across the board, while the untreated group either stayed the same or declined.[17]

While these results were striking, this study had many flaws, including the sample size (297 kids), and the self-selection for a surgical intervention (the families chose to have the surgery—the kids were not randomly selected). This is why Gozal and his colleagues did not stop at this one study. They acknowledged a link, not a cause, between sleep-disordered breathing and poor school performance. Because their data was statistically significant, they could state that the results were not due to chance. However, they did not fill the minds of the lay public with the idea that they had discovered a cause of poor school performance, aiming to put even the best of tutors out of business in favor of a good night's sleep. On the contrary, their findings simply led to more research, larger studies, and longer-term follow-up.[18]

Further investigation found a correlation between sleep-disordered breathing and neurocognitive deficits, including ADHD symptoms. Still, they did not claim that sleep-disordered breathing *caused* ADHD. The following decade or so has seen numerous research studies and anecdotal reports of a relationship between sleep-disordered breathing and attentional, academic, and behavioral deficits, and their significant improvement with better sleep. But again, a correlation, link, and even association. But not a cause. And perhaps just as important, not a cure. Sleep-disordered breathing does not cause ADHD; its resolution does not cure it. Sleep-disordered breathing may magnify symptoms, and treatment may minimize them. A critical distinction for those seeking improvement versus resolution. Alleviating sleep disorders will lead to substantial reductions in ADHD behaviors such as inattention, poor focus, and difficulty with impulse control.

Did the increase of ADHD have anything to do with the increase in diagnoses of sleep-disordered breathing? Probably not. Kids have been snoring for years. The recognition of a problem does not mean the problem hadn't been there to begin with. Sleep disorders' known correlation to adult issues such as daytime fatigue, dangerous driving, and poor work performance far preceded the recognition of the implications of pediatric sleep disorders. This does not make the latter a new problem; just a newly recognized one.

In a contrary example, Dr. Andrew Wakefield's presentation of a handful of children (twelve) exposed to the MMR vaccine showed a loose association with onset of gastrointestinal disorders and pervasive developmental disorder, later known as PDD and even later known as autism. This was misconstrued as evidence that vaccines *cause* autism. But the truth is vaccines are not even loosely associated with causing autism, and this has been demonstrated in countless studies of thousands and thousands of children. Authors of one study found no causal relationship between the MMR vaccine and autism based on reviews of thousands of children: those who received the MMR vaccine and those who did not had the same incidence of autism, with no temporal relationship between receiving the MMR vaccine and autism diagnosis. Studies similar to this large population study looking at hundreds of thousands of children over many years, assessing not just MMR, but multiple vaccines, also found no link, association, or cause between vaccination and autism.[19] This topic, which is clearly near and dear to my heart, will be a subject of Chapter 10. But suffice it to say that vaccines don't cause autism. The illnesses they prevent, on the other hand, can have deadly consequences.

Who doesn't love blueberries? They are, for the most part, sweet and they go with just about everything. Blueberries are often touted as being a "superfood" because they are rich in flavonoids, compounds that possess antioxidant and anti-inflammatory properties. So when a study shows that they also improve brain function, pass the blueberry pancakes, please. But headlines such as "Blueberry Concentrate Improves Brain Function in Older People" deserve a closer look. While this was a double-blind, prospective study, it included only twenty-six people, twelve of whom were given high-dose daily blueberry concentrate for twelve weeks; and fourteen of whom ate their normal blueberry-deprived diet (although it's possible these fourteen had lots of regular blueberries in their diets, they didn't get the concentrate that their twelve lucky counterparts did). The investigators then looked at MRI scans and did some memory cognition testing in all subjects and found that those who ate blueberry concentrate had more blood flow to memory centers on their scans and performed better on memory tests. This information may be valid. It had more than double the number of subjects of Wakefield's study. It was also sponsored by CherryActive, Ltd, makers of, you guessed it, blueberry extract concentrate.[20]

As the line in a Joe Jackson song goes, "Everything gives you cancer." Thankfully, this cynical line is far from the truth, but it does point out that, even back several decades ago when this song was written, we were bombarded daily with scares about cancer-causing chemicals, habits, and predispositions. Indeed, some substances, habits, and predispositions *do* cause cancer or at least increase our chances of developing cancer. But we must be wary of fast claims, as some studies are done in test tubes, using exponentially more of a substance than one would ever be exposed to over a lifetime, or they are performed in animals, such as fruit flies, smaller than a comma on this page. And remember that *cause, association,* and *link* carry three drastically differing meanings. Cigarettes cause lung cancer. Reading this is associated with being smarter. Not reading this is linked to missing the big picture on cause, association, and link.

HYPE ALERT

* Links and correlations are the same, but to be a *cause* entails 100 percent correlation.

* X can cause Y in a laboratory setting, but have no significant health impact in humans.

* A "healthy" ingredient can be toxic in high doses (e.g., fluoride, vitamins, omega-3 fish oil, and alcohol).

* Pears contain more formaldehyde than any vaccine.

* Labels can be misleading; *unscented, fragrance-free,* and *natural* products can still contain harmful ingredients.

4

GET ME OFF YOUR F*CKING MAILING LIST: A STUDY WORTHY OF YOUR ATTENTION

How to Make Sense of Medical Research Jargon

IS EVERY PUBLISHED STUDY MEANINGFUL AND TRUSTWORTHY?

HOW CAN YOU DETECT A BAD, HYPED STUDY?

WHAT DOES *DOCTOR RECOMMENDED* MEAN?

IS THERE A DIFFERENCE BETWEEN *CLINICALLY PROVEN* AND *SCIENTIFICALLY PROVEN?*

In 2005, computer science professors David Mazières of Stanford and Eddie Kohler of Harvard wrote a fake ten-page paper called "Get Me Off Your Fucking Mailing List."[1] It was a joke to reply to unwanted conference invitations. Their paper—which contains only those seven words, repeated 863 times—has been in circulation ever since, and you can easily access it online. In 2014, another computer science professor forwarded the bogus paper to the *International Journal of Advanced Computer Technology*. Apparently, this professor, Peter Vamplew of Australia, was annoyed at unsolicited emails from the journal (as well as others) and sent the old paper in retaliation and in hopes that the spamming would stop. Never did he expect what happened next: the paper was accepted for publication and rated "excellent." Despite sounding awfully distinguished, this publication is a predatory open-access journal that spams scientists by offering to publish their work if they pay a fee (Vamplew declined the "offer" to pay $150). By definition, open-access journals are available for free online (the majority of journals, especially the prestigious ones, are closed access and require a subscription of some sort or payment to download papers).

While the tongue-in-cheek "Get Me Off Your Fucking Mailing List" debacle gives us all a chuckle at how ridiculous this pseudoscience world of fraud can be, what's scarier is when articles that seem legitimate but are actually just as nonsensical are accepted to scam journals. In 2013, *Science* writer John Bohannon created a fake name, and a completely fake study, which anybody over the age of fourteen would recognize as parody. His study claimed that a substance in lichen showed cancer-fighting properties. The methodology was off, the data were bizarrely presented, and the results made no sense. The made-up author and coauthors did not exist, and the institution from which they submitted was made up, too. Bohannon submitted his paper to 304 open-access journals, and 157 of them accepted it for publication.[2]

This sting operation left many open-access journals needing to answer some questions, apologize, or shut down. Certainly a start. Another sting operation involved a made-up professor asking to be on the editorial board of multiple open-access predatory journals. The pseudoprofessor invented a Polish name that literally meant "fraud." She was invited to be on editorial boards, or even to be an editor in chief, by multiple journals. One even promised that the position would require no work. Busted again.[3]

Predatory journals have long been a problem in the scientific community. Because the burden lies on scientists to "publish or perish" in a world where gaining acceptance to prestigious journals is challenging, a dark, unscrupulous market that lacks standards is somewhat inevitable. The open-access model has led to a great many new online publishers. But many of these are corrupt and exist only to profit from authors who pay for their papers to be published (sometimes for thousands of dollars). The number of predatory publishers in 2011 was only 18, but by 2016, there were 882.[4] It gets worse. Journals are now getting hijacked: Stand-alone publications (predatory or otherwise) are being stolen; people counterfeit the website, solicit journal submissions, and take the author fees directly on submission. Over a hundred of these exist at this writing, a number that will likely continue to grow. In 2016, the U.S. Federal Trade Commission began to recognize this ever-growing problem by suing some of the larger scam publishing conglomerates.[5] Even the high-level journals can be fooled. And high-level researchers can be fooled by bottom-dweller journals.

FAKE NEWS AND FALSIFIED STUDIES

How can you spot scientific red flags in the media and the headlines that tout these fake studies? Is there a difference between the *New England Journal of Medicine* and *World Journal of Science and Technology*? You bet. But even the most legitimate medical studies can be wrong; they are often biased and flawed in unique ways or reveal findings from a handful of people rather than a group numbering in the thousands or, preferably, tens of thousands. Perhaps the most notorious journal article in one of the most highly regarded journals was published in *The Lancet* in 1998.[6] Andrew Wakefield, a pediatric gastroenterologist I mentioned in the previous chapter, aimed to link the use of the measles/mumps/rubella (MMR) vaccine with what was then called pervasive developmental disorder, commonly known today as autism. As I called out, this study included a scant twelve children, which should have been a tip-off from the start. Alas, even *The Lancet* was fooled, and this article became the basis of an international debate on vaccine safety, which continues today. Despite retraction of the article due to falsely reported data, Wakefield's publication remains a landmark study in many ways. If the editors at *The Lancet* could be fooled, how can even the most medically informed know the difference between aboveboard research and a scam?

A highly regarded service of the *British Medical Journal* that vets new studies for clinicians finds that on average 6 percent of all new journal articles published each year are good enough to mean anything. Which means a breathtaking 94 percent of studies published are not well designed and relevant enough to inform patient care or to lead to any changes in how we take care of patients. In their conclusion, the authors write, "Exaggeration in news is strongly associated with exaggeration in press releases."[7] Moreover, one study's conclusion will be totally opposite to another study's bottom line. One day coffee and eggs are good for you; the next, not so much. What's the truth? Most well-designed studies have a kernel of truth, so the truth lies in the sum of all the research, which can be hard for anyone to evaluate closely and arrive at a sane conclusion amid competing data.

So how can we as health consumers make sense of all this contradiction and confusion? Let's start with a basic question: What is a "study"? First, there is the case study, or case report, for instance: A previously healthy thirty-year-old goes to her doctor. She has had three weeks of fevers, diarrhea, and a strange rash. Her hair has also begun to fall out. She undergoes multiple blood tests, X-rays, and consultations with specialists. She had traveled to a remote tropical island four weeks prior and was bitten by a strange mosquito carrying a strange virus. Fascinating! Let's write it up as a case study. It's scientific, it's real, and it happened! *Once.* Yes, that is an absolutely valid scientific paper and is considered a bona fide study. Now what if there were two or even three such patients? Science again. This time, we can call it a case series.

This is not to say that all case reports, case series, or even survey studies are nonsense. Most are not, and most lend genuine value to the medical literature and can be considered good science. However, one must be critical in what one reads or hears, even if buzzwords such as *science* and *research* are thrown in. As physicians and researchers, we are accustomed to automatically assessing quality research—is it published in a peer-reviewed journal, or a journal that invites authors and requires fees for publication? Do the authors have a conflict of interest? In other words, were the authors testing a drug and being paid by that drug's manufacturer? While this is not a reason, by itself, to discredit a research study, the authors gain more credibility if they provide full disclosure of any and all current and past conflicts of interest. All good journals require this, with serious penalty if this is not provided honestly. Readers can recognize the differences in quality in journal articles, whether they read the articles themselves or see them cited in other media.

I chew a particular brand of sugarless gum—let's call it Acme Gum. Why wouldn't I? Four out of five dentists surveyed recommend sugarless gum for their patients who chew gum. Well, I know five dentists. Maybe four of them recommend sugarless gum. But maybe I'm the only patient in my dentist's practice who chews gum. Maybe these four dentists aren't recommending Acme Gum, but another sugarless gum. Or maybe Acme is paying the dentists big bucks to recommend it. Or maybe the spouse of the dentist works for Acme. Maybe the dentists don't recommend chewing gum at all, but if a patient insists on it, they'll recommend that it be sugarless. Or maybe, just maybe, researchers at Acme Gum did a blinded survey of all 150,000 members of the American Dental Association, asking, "What gum do you recommend to your patients who chew gum?," received an overwhelming 90 percent response rate, and of the 135,000 respondents, 108,000 recommended Acme. More likely it was five handpicked dentists. And what brand does that fifth dentist recommend? Sugar-full Gum? Or was that fifth dentist so annoyed by the bizarre question that he wouldn't recommend any brand at all, but perhaps recommended more dental visits to the gum chewers?

Survey studies are notoriously challenging. I've been on the surveyor as well as the respondent end. I've queried about specific surgical practices to all members of a given specialty (thousands) down to members of a subspecialty (hundreds). A survey study can get as little as a 15 percent response rate or even less, weakening the data tremendously. Answers to questions are rarely labeled yes or no, but more likely have choices such as always, sometimes, never. Or strongly agree, agree, moderately agree, moderately disagree, disagree, or strongly disagree. It's mind-numbing to participate in a survey, and it's even more mind-numbing to evaluate the data. That's why the dentist study is so beautiful. It was purely a marketing angle, required no real study, and it worked. When you read about or hear about a survey study, such as the gum angle, it's hard to know what it means. If it was a group of doctors, how were these doctors picked? Of those invited to participate, what percentage participated, and why? Were they paid to do so? Some surveys offer entry to a raffle for participating. Was the coveted prize worth it? Is that coercion, bribery, or just incentive? Did all participants disclose conflicts of interest? And just as most of us have answered surveys of all kinds, the answers are usually based on opinion, not results. Be suspect when a headline is based on a survey study. Unless you're a gum chewer like me. Or one of those four dentists. Then it's all fact.

REAL NEWS ON STEROIDS

Unlike me, my husband is an early riser and combs through multiple newspapers every day long before the sun comes up. He's the consummate surgeon, and my article curator, able to weed out the fluff and get to the important stuff. As I sit bleary-eyed with my black coffee on dark winter mornings, he often hands me several articles with red-penned circles around the titles from one or more of the three print newspapers we receive each day. "Here. Read this before your clinic today. Your patients will want some explanations." Or "Your cases may get canceled today. Read this so you'll know why."

A new study shows that computed tomography (CT) scans cause cancer. A new study shows that ear tubes to either alleviate recurrent infections and/or chronic ear fluid are unnecessary. Michael Jackson died from an overdose of propofol, an anesthesia drug that kills. A young girl had a horrendous postoperative bleed after tonsil and airway surgery and is now declared brain-dead.

These were all real news, real studies, and real case reports and had real, albeit temporary, impacts. But as with most headlines, one must dig a little deeper. Indeed, CT scans emit potentially harmful radiation, which may have long-term impacts on those receiving multiple scans over a lifetime. A retrospective cohort study about exposure to CT scans increasing one's risk of cancer was performed in the United Kingdom and published in *The Lancet* in 2012. A *retrospective* study means that the authors looked into the past, and *cohort* means that at least two groups were studied. The researchers reviewed data on cancer patients under age twenty-two between the years 1985 and 2002—that's a long time in medical-study parlance. Specifically, they looked at the incidence of brain tumors and leukemia, as these cancers are most likely to be related to radiation exposure. They looked at nearly two hundred thousand patients. Big number. But the detail is that these patients had a prior cancer diagnosis, which could be why they were undergoing multiple CT scans. These patients had an increased risk of developing a brain tumor or leukemia, and this risk would increase with increasing dosing (equal to number) of CT scans. This risk is real, but here's how the numbers translate: in the ten years after the first scan for patients under age ten, one excess case of leukemia and one excess case of brain tumor per ten thousand CT scans was estimated.[8]

One excess case of either of these malignancies is one case too many. But one must look at the other issues—why were these CT scans ordered? Was it concern for a life-threatening bleed in the brain after an injury or a surgery? For a possible blocked intestine, requiring urgent surgery? How many of these ten thousand CT scans *saved* lives, and how many lives were saved by this assessment? We as physicians took this study seriously, as there were big numbers, and subsequent studies found similar results. However, simply stating that "CT scans cause cancer" is erroneous.

Ear-tube surgery is the most common surgical procedure performed in the United States in children. Close to 1 million of these one-millimeter spool-like plastic objects are placed into wee ears annually. About three hundred are done by me every year. Over my career, I've done a good ten thousand. Ear tubes act as temporary drainage systems for little kids who can't equalize pressure after recovering from a cold, whereby they suffer from either recurrent acute ear infections and/or chronic fluid that gets stuck behind the eardrums, leading to temporary hearing deficits, balance issues, speech delay, and overall misery for the children and anyone around them. Ear fluid can cause up to 40 percent hearing loss, potentially leading to speech delay and language deficits. Ear-tube placement immediately reverses the hearing loss, as well as the chronic discomfort from ear infections. To many, it's like a tiny miracle surgery. But miracles aside, would this fluid have *miraculously* dissipated on its own? The answer is yes. Sometimes. Ear-tube surgery is performed in a procedure room or operating room, the children are under anesthesia for about ten minutes, and they are able to go about their routine later that day or the next day. It is probably the most satisfying procedure we perform—it has little or no risk, requires little or no maintenance on the part of the parent or child, and is life-changing. But how can one determine if too many ear-tube surgeries are being done given such a high satisfaction rate, few long-term downsides, and so few negative outcomes? Well, some darn good research shows that ear-tube placement has become overkill, and that we're doing too many unnecessarily.

When one of those headlines landed near my coffee one morning, I knew I better get the data right before I saw my first patient of the day. Indeed, the headline stated, "Ear Infection? Think Twice Before Inserting a Tube."[9] Once I was caffeinated, this one I needed to address. This was not based on look backs, surveys, or a case report. One of the most world-renowned pediatricians and clinician scientists, Dr. Jack Paradise, sought to prove the surgeons wrong. He looked at thousands and thousands of children *prospectively,* from early toddler years to fourth grade and beyond. He studied groups of children who had either recurrent acute infections or chronic fluid behind their eardrums, with or without hearing loss. He followed these kids for years. He didn't need to look back; he looked ahead. He found that the kids who weren't treated (with ear tubes) right away ended up at the same level in speech and language down the line as kids who had ear tubes right away. The take home: doctors should not be rushing their patients to the operating room. The kids will do fine either way. But here's where one needs to read between the lines, and what's not written in the headlines.

First of all, just because there was a lag in ear-tube placement doesn't mean that the surgery was forgone altogether. It just means that if a child had fluid for five months instead of three months, the child would fare equally well in mid–elementary school. What is also too-often overlooked is the misery that these kids suffer from painful ear infections, requiring multiple courses of antibiotics, doctor visits, and sleepless nights. Also overlooked is parental strife—the worry when a child is suffering, the concern when a child's hearing diminishes, albeit reversibly, and the challenges of getting to doctor visits, of unscheduled trips to urgent care or emergency rooms, of canceled travel plans, and of loss of work time. The (extremely) valid point of the study was that doctors needed to create more stringent guidelines for ear-tube placement, not that they shouldn't be done at all. Guidelines were created, and most doctors have changed their practice based on these changes. But not based on the headlines.

Michael Jackson was pronounced dead in my hospital. This is no secret. I am not violating privacy laws. It was on the front page of every newspaper, the top newsfeed, and in every magazine that week. One of the biggest case reports in history. It was such startling and overwhelming news that Farrah Fawcett's death after a long illness, just a few miles away, a few hours earlier, was only a blip on the screen. The cause of Jackson's death was cardiac arrest, but why a fifty-year-old active entertainer would so suddenly die was called into question right away. It soon came out that he had received a drug (at home, no less) called propofol, an opaque white liquid intravenous anesthetic we like to call "milk of amnesia." He had been suffering from terrible insomnia, so was inappropriately given this drug to induce sleep. This was an insane idea for countless reasons—sleep under a general anesthetic is not the restful sleep one has naturally.

Propofol is a potent respiratory sedative that can result in your inability to breathe on your own, to name one of its side effects. This terrific drug anesthetizes the patient, reducing need for higher-risk drugs such as muscle paralytics, narcotics, and volatile anesthetic gases. Propofol has little to no hangover effect, and it rarely causes nausea or allergic reactions. It does, however, need to be administered by a trained provider in a fully monitored setting. This does not mean a cardiologist in a Bel Air living room. For weeks after Jackson's death, patients would refuse to have the "killer drug." Was propofol somehow safer before Jackson's death? And how did it become safe again, as I haven't been asked about propofol in years? The power of one tragic case report can be more than a study of thousands, but the lifetime of this power is short.

The pre-Christmas holidays are one of my busiest times of year. It's tonsillectomy season: the school-age kids have a nice long break to recover from their surgery, and it's the perfect time to use insurance after deductibles have been met. This was not the case in 2013. On December 9 of that year, Jahi McMath arrived at Children's Hospital Oakland to have a tonsillectomy as well as other related airway surgeries to alleviate her obstructive sleep apnea. Although the medical details of her case are as yet not known, she developed massive bleeding and airway blockage in the hours after surgery, leading to respiratory arrest. She was resuscitated, although tragically pronounced brain-dead by her doctors on December 12.[10] This led to one of the most publicized debates about brain death versus sustained life support in recent history, after the famous Terri Schiavo right-to-die legal case that ended in 2005 (if you don't remember: Terri was in an irreversible vegetative state and her husband wanted to take her off life support, while her parents did not. A highly publicized series of legal challenges ensued for years until, finally, her feeding tube was removed).

But perhaps more remarkable to both tonsil surgeons and tonsil patients was the unprecedented surgical outcome. This was a horrific case. A devastating case report. But so extraordinarily rare. It rightly called into question the indications and potential complications of tonsillectomy to all of us who perform the surgery. We had plenty of time in the following weeks to contemplate this, as most of our scheduled patients canceled their surgery.

While studies covering tens of thousands showing patients had excellent results following surgery do not make the headlines, one tragedy does. And just as the issue of the cause of Michael Jackson's demise faded into the background, so did that of Jahi McMath's. The focus of both of these cases became the legal struggles: Was Michael Jackson's doctor at fault? Is Jahi McMath still alive? Neither the dangers of propofol nor of post-tonsillectomy mortality are discussed anymore. Neither medical issue lasted long. A week or two after Jackson's death, all were again fine with propofol flowing through their veins. And by January 2014, my surgery schedule for tonsillectomies was overbooked.

No easy answer or code word enables one to discern shoddy research from the real deal. The first task is for the predatory journals to disappear. The academic system is known for putting the pressure on; a person's job status and career are tightly tied to quantity of publications, and people are willing to spend the money to get a paper out quickly, without recognizing when a journal is a hoax. Academic programs are becoming more aware of this dilemma and are beginning to focus more on quality and less on quantity. As a somewhat seasoned academician, I am always suspicious of any invitations for submissions, either for journals or for meeting presentations. While it's easy to feel honored by such invitations, these are more often than not scams. It gets blurrier because some of the open-access journals publish legitimate studies. But when they are asking the authors for up to $2,000 to have their work published, all in the name of rapid appearance and nearly guaranteed online visibility, most would (or should) think twice. The top journals accept as little as 5 percent of submitted articles. If a publication needs to solicit submissions, there's a problem. And timeliness should not be a factor. While routine articles can take up to twelve months or even longer to hit the press, even after acceptance to a journal, the timely articles are rushed to the head of the line and can be up on a journal's online site within a month. But that doesn't help the reader of even the most well-respected news sources. Journals titles can be highfalutin, reports may be from respected institutions, but it's easy to be fooled. But with a little bit of quick checking, you can also partake in some vetting of headlines.

First, check your first source. Was it a headline on your Facebook, Twitter, or Instagram feed? And if you click on the link, does the footer contain advertisements directly tied to the headline? (For instance, ads for ginkgo biloba extract after an article touting the health benefits of ginkgo biloba?). Look for some basic numbers: How many subjects were there, over how much time, and, when looking at cure, for how long did subjects remain "cured"? Watch out for data that's misinterpreted or somehow misrepresented to support a particular idea or theory. So when you read the headline "Many Studies Have Debunked the Saturated Fat Myth," read between the lines. How many studies exactly?

How many people were involved overall? What kinds of controls were used? Data manipulation is rampant.

Look for buzzwords such as *miraculous, groundbreaking,* and *remarkable*. For seasoned scientists, no study, no matter how big or small, is a miracle, breaks ground, or is remarkable. Your source should name the journal or meeting where the original study was published. While many of these require access codes to see the whole study, you should be able to publicly access the abstract, so you can read for yourself some of the details of the study, including objective, hypothesis, study methods, and conclusion. Start at PubMed.gov. PubMed, as it defines itself, "is a free search engine accessing primarily the MEDLINE database of references and abstracts on life sciences and biomedical topics. The United States National Library of Medicine (NLM) at the National Institutes of Health maintains the database."

If the abstract reads like an advertisement, it probably is. Most abstracts will also provide the type of study performed, and there are some basic ones to know. *Double-blinded* means that neither the subjects nor the investigators knew which treatment/intervention was received. *Single-blinded* means that either the subjects or the investigators knew the treatment/intervention. *Prospective* means looking forward in time, while *retrospective* means looking back on information obtained before a study was considered, also known as a *chart review* or *record review*. *Observational* implies no intervention (such as medicine, surgery, or diagnostic test), but simply observing subjects over time. *Interventional* means that something was done to the subjects—a diagnostic test, a drug administration, or surgical procedure. If none of this is provided in an abstract, or an abstract is nowhere to be found online, question your source.

Most health articles presented as widely visible news items have at least some validity. They were promoted either by a public relations department of a hospital or institution or received acclaim from several objective sources before being brought to the public. They tend to be large studies, involving years of background work and thousands of subjects. But even those can leave us in a quandary. Annual mammograms are recommended by most physicians, yet some solid, popularized data shows that they may not save lives and may even contribute to unnecessary breast surgeries, and future breast cancer risk due to radiation exposure (see Chapter 11). The PSA (prostate-specific antigen) blood test has been the mainstay in screening for prostate cancer, yet some large studies have shown that this test leads to many unnecessary prostate biopsies and surgeries, incurring more harm than good. This issue of conflicting data comes up all too often. Longtime recommendations from medical societies will be refuted, and this will make front-page headlines. Neither viewpoint is necessarily wrong or fallacious. Neither is necessarily "fake" or based on articles in scamming journals. It's actually a good thing that these questions come into play in both medical and lay forums. But they should do nothing more than lead you to raise questions, which is what we do as practitioners.

If a study shows a link between CT scans and future risk of cancer due to radiation exposure, this does not necessarily mean that there is a direct cause, but that the indications for and frequency of CT scans should be questioned. If mammograms are presented as potentially more harmful than lifesaving, you should ask your doctor, Do I really need one? Every year or every other year, or not at all? Even when large studies provide solid results, we still need to account for individual circumstances. The take-home for issues such as these is that they should raise questions, and this is a good thing. But even the best, most solid large-scale studies may give information and/or advice that is 180-degrees contrary to what you've been doing, and you may come full circle in the future. That's just the nature of medicine and science. All one can do is to step back, take in the information, and discuss it with a practitioner whom you trust.

CLINICALLY PROVEN

More disconcerting than flip-flopping recommendations are exaggerated or fallacious claims on products, with medical and research terminology thrown in, somehow putting them into the research-based level of credibility. Terms such as *doctors recommend, clinically proven,* and *studies show* can mean something substantive or absolutely nothing. In general, real studies don't apply these claims to results, as nothing in medicine is ever definitely clinically proven, even if a large study said it was. We all know that things can change down the line. Future studies will refute even the most valid present ones. After all, in years past, doctors not only recommended cigarette smoking as a source of relaxation, they were also *in* the cigarette ads. Doctors used to recommend that pregnant women should do nothing but rest. Now we know that is not the case.

The Livestrong website claims that certain weight-loss products are "clinically proven" to work because they "work in the clinical setting."[11] They even back their claims with journal articles. One article was the basis for selling green-tea-extract supplements for weight loss. This article, published in 2007 in *Obesity,* found increased weight loss, lower blood pressure, and reduction of LDL (one of the bad cholesterols) in people who used green-tea-extract supplements.[12] On closer look, one sees the study examined 240 Japanese who dieted for twelve weeks while half took a high dose with the active substance in green tea extract (catechins), and half took a lower dose. The high-dose group did show more weight loss, and this difference was "significant" (not just coincidence in medical terms), but still small. The differences in blood-pressure and cholesterol changes also were significant, yet small.

At the end of the article, the authors, who did not receive any funding for the project, state, "The costs of publication of this article were defrayed. . . . This article must, hereby, be marked as 'advertisement.'"

We are left wondering if this clinical trial was unbiased. Taking green-tea-extract supplements may, indeed, lead to increased weight loss during a diet. But we still don't know for sure based on these claims. (While I'm on the topic of nutrition studies, I should add that in general these studies have limits. It's difficult, if not impossible, to conduct traditional studies on people's diets using a randomized, controlled design as is done with pharmaceutical studies. Plus, foods contain a staggering number of different ingredients. If we find associations between a particular type of food or beverage and a health effect, such as weight loss, the exact components that produce such an effect are difficult or impossible to isolate given the complex composition of the foods and beverages—not to mention potential interactions among other ingredients and underlying genetic factors. And then we have the practical issues of basing a nutritional or weight-loss study on people's possibly-dishonest recordings of what they ate, as well as controlling for their lifestyle (e.g., exercise habits, nicotine use), which can factor into their health and weight-loss efforts.

When you read labels of any product that say *clinically proven,* all this means is that the product helped in a clinical setting. This could have been on thousands of occasions, hundreds of occasions, or just once. The same goes for *studies show, doctors recommend,* or *scientifically proven.* No outside agency regulates the use of these phrases. So if my husband and I (two medical doctors) recommend Cheetos for breakfast, the makers can slap *doctors recommend* on their label. Our kids would be thrilled.

A term that usually doesn't hit the airwaves or the print media is *P value,* but it does hit the hearts and minds of those of us in the research world. A P value is a predictive measure of probability of difference between two factors—either treatments, diseases, or as a result of any type of intervention. We measure P value as an indicator that a difference is due to the intervention, not due to chance. The lower the P value, the more likely a difference is to occur. In research arenas, a P value of less than 0.05 is where significance begins. This means if an event occurs one hundred times, the probability that the difference occurs because of the intervention, and not just by chance, is 95 percent. It gives a degree of certainty to the data.[13]

For instance, if you flip a coin, the P value of the difference between heads and tails will be 1.0. The probability is 100 percent that the outcome of heads will be by chance. But if the coin is weighted, favoring heads, the P value will go down, as the likelihood of heads is due to more than pure chance. The heavier the weight, the more likely the coin will land on heads, and the lower the P value. In the lay press, P values are rarely given, but you may see words such as *significant* or *different.*

But what we look for as words indicating actual value is *statistically significant,* which points to a low (good) P value.

There is little or no way that one can tease through the nonsense and the good that's put on our daily plate of health news. Studies may make headlines if a celebrity is suffering from that disease, if the political climate is attuned to a particular issue, or if something is so groundbreaking that it must be told. But the dogma of the way these studies are presented needs to be taken away. Our medical societies spend years, using countless studies and reviewers, to come up with recommended guidelines for evaluation, diagnosis, and treatment of a given entity, only to have those guidelines altered, either slightly or not so slightly, a year or decades later. While this is inherently frustrating, it is also the nature of medical progress. Nothing you read or hear, no matter how big the study, how large the database, or how respected the journal and researchers are, is the final word.

HYPE ALERT

* Predatory journals abound today; pay attention to and appreciate studies that involve large numbers of people—not just a handful.

* Data can be easily manipulated, so read carefully.

* Don't fall into the trap of thinking *doctor recommended* or *clinically proven* means anything or meets established standards.

5

TIPPING THE SCALE ON A BALANCED DIET: YOU ARE NOT ALWAYS WHAT YOU EAT

How to Filter Out the Noise on Juicing, Going Gluten-Free, Detoxing, and GMOs

IS THERE A ONE-SIZE-FITS-ALL "BEST" DIET?

WHY DO PEOPLE SWEAR BY "DETOX" PROTOCOLS?

DO "SUPERFOODS" REALLY EXIST?

IS GLUTEN THAT BAD—FOR EVERYONE?

WHAT DOES IT MEAN TO HAVE A FOOD ALLERGY?

Diets are becoming so exceptionally micromanaged that food is no longer the least bit enjoyable. How many of you have tried to avoid gluten? Tried a juice or "detox" cleanse? How many times have you listened to a weight-loss sales pitch in the media and rushed out to buy a supplement? Lists of allergies (not just one) are now taped to the walls of schoolrooms. At my son's preschool, the allergy lists included flaxseed, olives, goat's milk, and balsamic vinegar. I swear that my palate was not that experienced at age three, let alone five decades later.

When my daughter was young, her nose used to turn red and itchy after eating ketchup—until she went to a birthday party where ketchup was served and her nose stayed unchanged. When I checked our fridge,

I found out we had organic ketchup. So I went back to good old conventional Heinz 57, and she is Rudolph no longer. For a while, she would tell people she was "allergic to organic ketchup." I let her do this because I got a kick out of my kid's needing to eat the non-organic stuff to get by in this world of organics.

In recent years, we've seen celebrities, athletes, and even some respected, popular doctors come under fire for touting health and diet claims that may or may not have any connection to respectable studies. People love to see diet ideas that are out of the box, even when they come from people who seem unlikely touts or who veer from traditional medicine despite being a member of the licensed medical tribe.

The diet industry is colossal; Americans spend north of $60 billion annually to drop the pounds.[1] That's more than eleven times what we spend on cancer research each year. But more people struggle with weight than with cancer. Every New Year's Day, millions of Americans resolve to go on a diet. They say, "This is the year. This year I'm finally going to stick to my [fill in the blank: Paleo, Sugar Busters!, Weight Watchers, low-fat, low-carb, fast-metabolism, pescatarian, vegan, DASH, Jenny Craig, Zone, Atkins, South Beach] diet and lose the weight for good." Most of them fail.

Losing weight permanently isn't so easy, for if it were, then diet books wouldn't dominate the health genre and talk shows—including well-respected morning news programs—probably wouldn't be so popular (and often driven by the next diet/weight-loss/health-tip miracle). So what are the best diets, and do such diets exist? What do our top health institutions have to say about the diets that are the easiest on the body, mind, and wallet—the healthy ones we can stick to and that won't drain our bank accounts? And which ones are complete hype? Moreover, should the focus of dieting be on weight loss, or on being healthier and lowering our risk for illness?

These questions are on millions of people's minds and, when not explored safely, can lead to desperate, hyped measures: detoxes, juicing, and master cleanses; colonics; and diet pills and supplements marketed to speed up your metabolism. Many of the supplements sold to spur weight loss are unregulated. And regulated pharmaceuticals that promise to help you lose weight can come with serious side effects (e.g., one weight-loss drug's long list of potential side effects includes everything from liver damage and seizures to mania, depression, aggression, and thoughts of suicide). The lengths people will go to shed the pounds is borderline unreal. While some of these strategies might serve a function under certain circumstances, they are often not as "healthful" and weight-loss promoting as they are touted to be. And as I just illustrated, they frequently entail unwanted biological effects, or no effects at all.

But what about the positive psychological effects? For example, what if you feel better after you've experienced a detox diet? Any good outcomes of such treatments can probably be explained by pseudoscience's best friend: the placebo effect. I'll be going into depth about the placebo effect in Chapter 9. For now, let's turn to the rules to dieting safely.

RULE #1: PICK AND CHOOSE WHAT WORKS FOR YOU

When you contemplate dieting, the concept is best approached, and dieting ultimately most successful, if you think about "your diet" as opposed "being on a diet." But even with this angle, is there such a thing as "the best diet," or "the right diet" to maintain not only a healthy weight, but also a healthy body?[2]

It is no surprise that an unhealthy diet can lead not only to chronic obesity, but also to chronic illness and premature death. And the converse is true for a healthy diet. Plenty of studies show that continued poor dietary habits decrease life span, measured in numbers of years of life, and also decrease so-called healthspan, which is years of healthy life. Lifestyle-related chronic illnesses, with poor diet being one of the major players, are on the rise. Life expectancy is not going up, but chronic illness—beginning at younger and younger ages—is. Type 2 diabetes, which is largely driven by obesity and, before that, poor dietary choices leading to the obesity and metabolic disorder, was rarely seen in children several decades ago. The disease was considered an ailment of adults, particularly older adults (it used to be called adult-onset diabetes). But today, about 208,000 Americans under age twenty are estimated to have diagnosed type 2 diabetes.[3]

The list of doctor-approved diets is ever expanding, including, but not limited to, low-carbohydrate, low-fat, vegetarian, low-glycemic, Mediterranean, Dietary Approaches to Stop Hypertension (DASH), Diabetes Prevention Program (DPP), Paleolithic (Paleo), and vegan. And the list continues to grow. All of these diets have merits, as they all include healthful choices including fresh whole vegetables, lean meats, minimal to no processed foods, nuts, legumes, small amounts of alcohol, and minimal sugar. But among these is no "best" diet, as each has qualities that will or will not work with particular lifestyles and health conditions. It is certainly reasonable to pick and choose aspects from each of these approaches, without worrying about slapping a label on the type of diet one picks. Let's throw in a surgical analogy: One operation can be done ten different ways by ten different surgeons. Each way is the best way for that particular surgeon because it's what works for her, for her patients, and it's what she does regularly. When we have residents, I love that they question certain aspects of a technique and even ask my opinion about another surgeon's technique for the same operation. As I did for myself when I was training, I tell them that they have the luxury of seeing what they do and do not like about one operation being performed ten different ways. They should choose what they like about each technique and apply it in the future. Diets can be just like surgical techniques. Pick and choose what works. Now, there's evolution for you.

A simple caveat about the smorgasbord of diets is that it's not just about quality, but also about quantity. So many of the trendy diets emphasize the all-you-can-eat concept: "You have tea and juice three meals per day, but you can eat all of the vegetables you want! All day!" The all-you-can-eat idea, which hearkens back to many of our college days of ten-, fifteen-, or even twenty-pound weight gains, is so unhealthy. Even outside the college cafeteria, where meal cards often cover all you can eat, restaurants that offer all-you-can-eat salad bars, pasta refills, bottomless bread baskets, and even soda refills only contribute to the false need to eat it all. At one sitting. There is absolutely no reason, aside from the feeling of comfort and relieving stress and anxiety, to eat *all* you can eat. Even if the foods are low in calories, why does one feel compelled to stuff oneself to the gills just because it's allowed or comes with the price of admission?

RULE #2: DON'T BE FOOLED
BY SUPERFOODS

Is there such thing as a superfood? The term *superfood* has no medical meaning whatsoever. It is a marketing tool to describe food with supposed health benefits. In the late 1970s, spaghetti was out and pasta was in. Pasta was a superfood. Add some pesto to your penne, and you'd be the next "super" Food Network star. Bran muffins, with five hundred calories of carbs, sugar, and fat and a little bit of bran flakes, were a superfood of the eighties. And so were granola, granola bars, and frozen yogurt. These are all foods that touted some semblance of health benefit, but, for the most part, were and are *unhealthy*. What they were was trendy, hip, and eaten by the highly educated, health-conscious set of their day. They may not have been called superfoods, but they were the superfoods of the time. In reality, they contributed to the beginnings of the obesity epidemic—being packed with refined sugar, having low or no protein, and having low nutritional value. One classic *Seinfeld* episode was hilariously about Elaine's gaining weight from eating what she thought was low-fat frozen yogurt, when, all along, she had been duped into buying the full-fat treat. As if that would have made a difference—low fat or high fat, frozen yogurt is loaded with sugar and calories. And then there are the toppings. Oh, the toppings. The years following the fooled-ya fro-yo brought low-carb superfoods: tofu, Tofutti, soy products, and yam fries. Certainly all were healthier than sugar-filled bran bombs and what was essentially ice cream in yogurt's clothing, but each carried its own baggage when eaten to excess.

Remember those store-bought fruit platters, surrounded by that stiff green garnish that you'd push out of the way to get a perfectly sliced half-moon pineapple piece? Well, push no longer. There's your superfood of this millennium—kale. Which can cause hypothyroidism if eaten in excess. Kale, as well as other cruciferous vegetables such as broccoli and cauliflower, are what's known as goitrogenic. A goiter is an abnormally enlarged thyroid gland, and those with known thyroid disorders are at risk of developing more thyroid problems when such vegetables are eaten to excess. How much is too much? Well, this has never been assessed, as consumption needs to be inordinately large to cause a problem. We're not rabbits, after all! But the juicing craze has led to just that—inordinate amounts of these vegetables consumed in an eight-ounce glass. A huge bag of kale, good enough for a week of salads in our house, can be juiced down to a mere four ounces of green liquid. And when it's blended into a gnarly green smoothie, one of kale's prime benefits—fiber—is lost. Mixed with your kale, don't forget your berries. These are longtime well-known fruits that one would enjoy seasonally or on top of frozen yogurt or mixed into granola. But they are also superfoods. No major organ damage results from eating too many berries; the only possible negative effects would be too much sugar intake and some loose stools. Black licorice, another short-lived superfood, has antioxidant properties, but overeating it can lead to hypokalemia (low potassium) and cause heart arrhythmias. How much licorice (the culprit is the substance glycyrrhizin in natural licorice) is too much is not clear. The good news is that most candies marketed as licorice have the fake stuff or, at most, just minute amounts of licorice extract, and only in the black licorice variety, not the red.[4] So not to worry—sit back and enjoy your movie while you eat your red "licorice."

In 2017, a new study out of Finland (published in the *American Journal of Epidemiology*) warned pregnant women against consuming too much licorice after researchers at the University of Helsinki discovered that its natural sweetener—glycyrrhizin—can have long-term harmful effects on the development of the fetus.[6] In the study, kids around the age of thirteen who had been exposed to large amounts of licorice in utero were outperformed by others in cognitive reasoning tests carried out by a psychologist. The difference was the equivalent of about seven IQ points. According to the parents, these kids also had more ADHD-type challenges and the girls entered puberty earlier. Granted, other variables could be at work here for these outcomes, and the study was small and further research is warranted. But it's interesting nonetheless. (For the curious: Glycyrrhizin amplifies the effects of the stress hormone cortisol by inhibiting the enzyme that inactivates cortisol. The development of a fetus requires cortisol, but in large amounts, it is harmful.)

At least we can all pronounce the superfoods of yesteryear. But what makes superfoods "super" these days is a challenging pronunciation. Those who say açaí, quinoa, and turmeric correctly set themselves apart as superfood connoisseurs. In all seriousness, the majority of these superfoods can be supergood for you. They are packed with nutrients and have little of the bad stuff such as simple sugars, fast-metabolizing carbohydrates, or too much fat. Eating blueberries, açaí berries, or quinoa is good for you. But will it reduce the incidence of illnesses such as cancer, diabetes, high blood pressure, or high cholesterol? The short answer is yes, but as with most health recommendations, they are not stand-alone preventions nor cures. Yes, they contain nutrients. Yes, some contain the evocative antioxidants, veritable lead shields from cancer's ugly face. Blueberries, for example, have a moderate amount of nutrients such as vitamin C, but their claim to fame is that they have antioxidants. They do contain the antioxidant anthocyanin; however, the antioxidant properties are only active in a petri dish, not when consumed. And consider the following: when whole fruits are pulverized in a blender as part of a juice, or juice cleanse, you lose that all-important fiber that helps control blood sugar.

Here's a perfect example: I love to drink Naked Juice. It tastes great, and it appeals to my sweet tooth. And of course I feel that illusion of health when I don't go for the can of soda. After a long day in a dehydrating operating room, having been on my feet all day, and subsisting on graham crackers and black coffee, Naked Juice hits the spot as I leave the hospital to head home. My favorite flavor is Mighty Mango. It packs 290 calories of carbohydrate and sugar, almost no protein, and no fiber. It does, however, contain enough vitamin A to support a small village where those mangoes are grown. And that's about it. Why is it naked? It contains no *added* sugar, but who needs added sugar when the fruits themselves are laden with it. One serving contains 1¼ mangoes, 1¾ apples, half an orange, a third of a banana, and a squirt of lemon. I think if I sat down and ate all those fruits as quickly as I drink my drink, I'd explode. Or at least feel full. And for good reason: as I just mentioned, one of the major benefits of fruit is its fiber. Fruits are meant to be chewed just enough to allow them to be swallowed, but not so much that they lose the fiber. What aids in satiety and digestion with these superfoods is that they are eaten and digested slowly. It's the bulk of the fiber that makes fruit healthy. The vitamins are a small added bonus. Squeezing supposed superfoods into bottles and jars takes the super quality away.

Kids are now being fed from food pouches. These are not infants, but kids with teeth. Are they in that much of a hurry that they can't eat food as a task unto itself? We evolved to have molars and complex chewing abilities to be able to eat solid food. Part of the juicing and meal-replacement ideas come from the perceived lack of time. There's no longer time to eat four ounces of yogurt, so we have Go-Gurt? The juicing fallacy lies in the juicing process itself. If you simply eat the whole fruits that are squeezed into a juiced product, you gain fiber, more satiety, and slower digestion, leading to delayed hunger. The juicing of fruits may leave the vitamins intact, but the fiber is destroyed. You have a mouth, a stomach, a small intestine, and a large intestine, which are better than any juicer on the market. And these internal juicers hold on to the bulk of the fruit's fibers, makes them last longer in your system and mechanically slow your digestion. Juices go through your digestive system quickly, so you only get the benefit of sugars and a little bit of vitamins. After my sugar-packed pseudo-superjuice treat, which I can finish on the way home if there's just enough traffic, I've worked up quite an appetite for a full dinner.

The new buzz on superfoods is antioxidant. While we all love oxygen, oxidation of certain molecules leads to free radicals, which can, in turn, damage cells and even lead to cellular changes and cancer. Substances such as vitamin C, vitamin E, and vitamin A are antioxidants, providing balance to oxidation by inhibiting such chemical reactions. But as with most scientific entities on a molecular level, the process does not directly transfer to dietary intake. No study had been made to document that dietary intake of antioxidants or antioxidant supplements prevents or treats cancer. In fact, some antioxidants, notably beta-carotene, have been found to *increase* cancer-related mortality as opposed to decreasing it.[6]

RULE #3: APPRECIATE YOUR BODY'S OWN DETOX CENTERS

Another hot buzzword, which used to be limited to those who were undergoing rehabilitation from abuse of various substances such as alcohol, heroin, or cocaine, is *detox*. In detoxing from such substances, one either goes cold turkey, stopping use of the drug altogether, or gradually decreases use of the drug by substituting a similar substance. Cold-turkey detox is tough. Alcoholics undergoing detox often need close medical monitoring, as they can develop delirium tremens, or d.t.'s, leading to seizures and even death. In many hospitals, if an alcoholic is admitted for any reason, including a health problem unrelated to alcohol abuse, he or she will receive medication to prevent d.t.'s. Some may require intravenous alcohol to prevent this, including during or right after a surgery. Patients cutting out heroin or other narcotics are also at risk for life-threatening complications if they stop abruptly. Many receive a long-term treatment plan that includes other forms of narcotics to get them off the drugs safely. Cocaine abuse, followed by suddenly stopping, can lead to life-threatening heart problems. Most who are undergoing detox do so in a detox center, either in a hospital or other treatment center.

Now to the good news. Assuming we do not have one of the exceedingly rare genetic metabolic disorders one only encounters in tertiary-care facilities, and assuming we are not trying to wean ourselves from substance abuse, we are all born with detox centers! They are called our livers, spleens, kidneys, sweat glands, and gastrointestinal tracts. According to popular lore, we are now all filled with toxins, and our body's natural detox centers are no longer sufficient to rid us of all of these evil humours. So here come the detox diets—juicing detox, detox cleanses, etc. The goal of these is to get rid of toxins contributing to anything from depression, fatigue, digestive disorders, obesity, and dull complexions to just general aches and pains. Some diets claim to accomplish this by cutting out all solid foods, drinking specific juices, and seeing how one feels after a few weeks. Well, if people cut out caffeine, sugary treats such as doughnuts and cookies, and fried foods for a few weeks, they would most likely feel pretty good—at least after they'd recovered from the coma of decaffeination and the feelings of starvation for lack of biting into that cruller in the break room at work. But "detoxed"? What toxins are we getting out of our system?

Our kidneys are great detox centers. Without them, we either need to be on a dialysis machine (which does the work of the kidneys, usually for three to four hours, three to four times per week), need a kidney transplant, or will die. Sounds like the kidneys rid us of some pretty important toxins, and not just the ones in our Doritos. The kidneys filter the blood, and the "toxins" in our blood are turned into urine, which we pee out. Free detox! The liver is also pretty critical. Without our liver, or a liver transplant, we die. Must be pretty important for detoxing, too. Livers filter out the "toxins" in the food we eat. Thanks, liver! The spleen, that stealth organ that even medical students are puzzled by, filters out the contaminants in our blood. Let's hear it for the spleen! And probably the most useful detoxifying organ is our intestinal tract. This twenty-five-plus-feet-long tube is simultaneously digesting food by breaking down the solids to semisolids (hey, it's a juicer, too!), as well as absorbing the good stuff and ridding us of the bad stuff, better known as poop. Detox! No sweat! But speaking of sweat, our sweat glands are present throughout our largest organ, the skin. Through our sweat, "toxins" such as ammonia, extra salts, and even extra water are let go into the air. Sweat may be smelly and unseemly, but sweating is a genuinely natural detoxification mechanism to regulate the body's fluid/electrolyte balance. It signals us to replenish, but it also rids us of unnecessary fluids and salts. An added bonus is that it regulates internal body temperature in hot weather or during exertion. Detox during exercise. No study has yet shown that detox diets have any benefits, but many show that our body's own detox centers are hard to match.[7]

RULE #4: DON'T BE TERRIFIED OF GLUTEN IF YOU DON'T HAVE CELIAC

Should you go gluten-free? Maybe.

Indeed there is science behind this trendy diet of today, coupled with the rising incidence of celiac, an autoimmune disease related to gluten—it's more than four times as common as it was fifty years ago. We must ask, How could an ingredient that has been in our food for thousands of years suddenly be so threatening? Why do some people get celiac while others don't? And why is it so much more common among Caucasians?

Gluten is a not a carb, as many think, given its link to breads and pasta. It's actually a group of proteins known as prolamins and glutelins. These proteins are viscoelastic, meaning they are stretchy and sticky. They make your bagel just right; they make your pasta just perfect. Breads and pastas are not known for protein content, but the majority of protein in breads and pastas is from gluten. Gluten is mainly seen in these foods, but is also hidden in others, such as soy sauce, ketchup, beer, and even ice cream.[9] We've been eating gluten for eons, but recent decades have given it such bad press that perhaps it will soon be denoted as a toxin.

To some, gluten is, indeed, toxic. These individuals have celiac disease, a genetic autoimmune disorder, now known to be linked with the genes HLA-DQ2 and HLA-DQ8. Interestingly, HLA-DQ2 is also linked to type 1 diabetes (formerly known as juvenile diabetes). This is not the diabetes we are seeing in increasing numbers with our increasing waistlines—this is the genetic form of diabetes, often found in young children or even infants. Many people who are diagnosed with celiac disease go on to develop type 1 diabetes, which needs to be managed with both diet and injectable insulin, not by diet alone. Classic celiac disease, seen in about 1–2 percent of the population, can present with pretty nondescript symptoms—abdominal bloating, discomfort, diarrhea, and loss of appetite when gluten-containing foods are eaten.[10] When, especially in kids, this then leads to malabsorption and resultant growth delays, further investigation to identify the presence of celiac entails both genetic testing and intestinal biopsies.

Dietary changes, removing gluten from the diet, are lifesaving. Going gluten-free is no joke for this population—it's not a fad, a lifestyle choice, or a way to lose weight. Gluten ingestion leads to dangerous malabsorption of fats, carbohydrates, and vitamins, leading to nutritional deficiencies, growth delays, anemia due to lack of iron absorption, and even bleeding disorders due to lack of vitamin K.

Gluten sensitivity is a different story. People with nonceliac gluten sensitivity will have similar complaints as those with celiac disease—abdominal bloating, discomfort, diarrhea. The mechanism of action of this, however, is different, as this population does not have antibodies reacting to gluten—so no damage is done, except for some discomfort. These gluten sensitives will remove gluten from their diet and feel better. So do those who go on high-protein/low-carb diets such as Atkins. Breads, cookies, and pastas are bloating. Eating these carb-laden treats is likened to eating sponges—they absorb liquids, raise blood sugar quickly, then drop blood sugar precipitously, not too different from simple sugars in candy. But gluten-free diets for weight loss or supposed sensitivity become a slippery slope. Most foods that would normally contain gluten, such as breads and pastas, now have gluten-free substitutes. They have a different texture, but I've tried them and gotten to like them. Why? The added oils, rice flour, and even salt or sugar! Gluten-free pretzels taste fried; gluten-free pasta is dense and firm. No wonder it's a popular diet. And while celiac is known to be associated with the inherited form of diabetes (type 1), the rise in gluten-free diets has led to an uptick in type 2 diabetes (linked to increased carbohydrate and sugar consumption).[11] More gluten-free food consumption has led to a relatively higher sugar consumption, as simple, refined carbohydrates have been substituted for complex starchy ones.

Another possibility is that gluten sensitivity is a form of IBS, or irritable bowel syndrome. IBS presents similarly to nonceliac gluten sensitivity—abdominal bloating, pain, malabsorption, diarrhea, and/or constipation upon eating certain foods. This food group is now known as FODMAP, which stands for *f*ermentable *o*ligo-, *d*i-, *mono*-*s*accharides *a*nd *p*olyols.[12] These include sugar-like substances, as well as sugar alcohols, included in foods such as artichokes, stone fruits, certain breads, dried fruits, and artificial sugars such as mannitol and sorbitol. Elimination of FODMAP foods can reduce symptoms of IBS.

We don't have all the answers yet; scientists have a ways to go in their understanding of both celiac disease and people who do not have celiac but who claim that they are sensitive to gluten. I'll also entertain the possibility that changes in our microbiome—the gut bacteria that collaborate with our physiology—may have something to do with it.[13] One interesting finding was that, just as introducing peanut products earlier may reduce peanut allergies, introducing gluten-containing food (wheat, barley, and rye) may reduce the risk of gluten sensitivities, especially in infants with a family history of celiac disease. While celiac disease is on the rise, what about the large contingent who claim to be gluten intolerant, or those who swear they are better off by avoiding gluten even without being gluten intolerant or having celiac disease?

Part of the rage for going gluten-free can be attributed to (or blamed on) celebrity claims that gluten-free dieting has left them feeling better, thinner, and healthier. Modifying gluten intake may have some benefit, but this may be simply due to some of the other ingredients in gluten-packed foods such as starchy breads and pastas. Substituting gluten-free foods for gluten-rich foods may not be the best answer. You might lose some of the gluten, but gain some unneeded fats, sugars, and calories.

RULE #5: KEEP A FOOD JOURNAL

The whole debate about gluten brings up a good question: What's the difference between a food sensitivity and a true allergy? And how can you know? While there is no scientific data to show that millions of people have become intolerant of or allergic to gluten, we do know that self-diagnosing one's dietary issues is, well, an issue. Those diagnoses are almost always wrong, especially when they creep across all of society. As I've stated previously, many people prefer to rely on anecdotes and stories rather than real data.[14]

A good case in point: over the past fifty years, monosodium glutamate, or MSG, has been demonized. Many refer to its alleged effects, from headaches and migraines, to heart palpitations, as "Chinese-restaurant syndrome," which came from a letter published in *The New England Journal of Medicine* way back in 1968.[15] Here's the kicker: we have no evidence that MSG causes those symptoms or any others, despite decades of study. And here's another surprising kicker: there are no chemical differences between the glutamate ions in the MSG we consume and the naturally occurring glutamate ions in our bodies. Contrary to what most people think, MSG is also found naturally in many other foods such as tomatoes (and tomato juice), peas, Parmesan and Roquefort cheese, potatoes, grapes, and mushrooms.

Similarly, wine must carry the "contains sulfites" label for reasons driven by politics, not science. A glass of wine contains roughly 10 mg of sulfites; two ounces of dried apricots, 112 mg. Wine-related allergies don't actually exist. Such additives, however, have been shown to correlate with migraine headache triggers. Many with chronic migraines will go on strict diets, avoiding "migranogenic" foods such as wine, chocolate, most cheeses, and foods containing soy sauce. While these foods are not necessarily causes of migraines in and of themselves, they can certainly contribute to their onset. But even those who suffer from migraines due to these substances are not allergic to them. But call it what you will—allergy, sensitivity, or aversion—what comes from these ails are products to combat them and get you to overcome your allergy, sensitivity, aversion—or pseudo allergy, sensitivity, intolerance, or aversion—and buy the wine. Products such as Üllo act as food-grade polymer filters to, you guessed it, remove sulfites from your wine.[16] The idea is nifty—a disposable filter is placed on the wineglass, you pour the wine over it, and voilà, sulfite-free wine. Well, sort of. It claims to "restore wine to its natural state." What they don't state is that the natural yeast contains sulfites. If you are one of the exceedingly rare individuals with a sulfite allergy, even the sulfites in the natural yeast will make you sick. Overall, Üllo is a decent idea, even if you didn't need it in the first place. Other population groups have a real difficulty in handling their alcohol—be it wine, beer, or liquor. Asians, in particular, carry lower levels of the enzyme ADH, or alcohol dehydrogenase, which is responsible for metabolizing alcohol into its breakdown products. Having less of this enzyme means that when you drink alcohol, it remains in its "pure" form for longer, which can result in skin flushing, vomiting, and even just being more intoxicated.

Despite what you think you might be allergic to, sensitive to, or intolerant of, you may be surprised at what's really making your stomach turn. To be in touch with symptoms related to food consumption, consider keeping a detailed food diary. Continue to eat your regular diet, without removing what you are convinced are the culprits (gluten, wine, coffee, dried fruits) for your (headaches, stomachaches, hunger, fatigue). Write down what you eat at each meal and/or snack, and the symptoms as they come. Be honest with yourself! You won't need to show this to your doctor, mate, dietitian, or your mother. The tricky part comes when the next time symptoms come, you remove *one* food item and see what happens. Again, be honest if things are no better or no worse. You might be surprised with the results.

RULE #6: UNDERSTAND TRUE ALLERGIES

If you're over the age of thirty, you likely remember that one kid with food allergies, but now food allergies seem to be more and more common each year. And it's true: food allergies among children have increased remarkably, some of which may be life threatening. Schools and day-care facilities are now nut-free, to avoid potential exposure for an allergic child. Before they enter kindergarten, many children learn to work their own emergency injector of epinephrine. School and day-care staff as well are now trained in use of the injectors.

Let's tackle one popular debate: Does eating peanuts and tree nuts during pregnancy make it more likely that your child will be allergic to nuts? No. In fact, a study that came out in 2015 debunks the theory that kids exposed to nuts as babies will go on to develop nut allergies. The new recommendation is to expose young children to nuts so they *don't* go on to develop allergies to them. In prior years, it wasn't recommended to give kids any nut products until age two at the earliest. Less allergenic products were recommended to be introduced first, such as cashew butter or almond butter. Studies showing that early exposure to nuts was better changed the recommendation to giving peanut products to kids as young as twelve months, or even younger, in efforts to reduce risk of allergies down the line. And higher maternal intake of peanuts, milk, and wheat products during pregnancy has been associated with a lower incidence of nut allergies, asthma, and allergic-related skin conditions respectively.[17]

Despite recommendations to begin nut products at two years, and then twelve months, neither recommendation seemed to stave off the increased incidence of life-threatening nut allergies, which continued to rise. So, more and more doctors started to try another tack for their patients: give nuts not only to pregnant women, but also to infants. New recommendations include feeding nut products (not whole nuts, as they are a choking hazard, but peanut butter or other nut butters mixed into cereal) to infants as young as four months old. These newer recommendations have already shown reduction in nut allergies in children.[18]

What's becoming almost as common as the life-threatening allergies, however, are the allergies that aren't. If a child with a history of rashes or upset stomach, for example, gets a blood test for food allergies, he or she may receive a multitude of false positive results. Many of these reactions occur in the lab, not in the person. When kids go for allergy testing, especially for food allergies, this often involves a series of blood tests. The blood is exposed to various food proteins. If the blood reacts to a given food protein, the child is labeled as allergic. The child may not, however, have the actual allergy when that food is eaten. The allergy lists posted at some of the most prestigious preschools is longer than most menus at restaurants. That one would know that a three-year-old is allergic to tofu, boysenberries, chickpeas, or pistachios—is that a sign that our privileged tots are getting gourmet palates decades before their parents had the opportunity to even taste such exotic foods?

Dairy has also become an interesting food group to shun. While many are lactose intolerant, few are truly allergic to dairy. Lactose is a disaccharide sugar in milk, which requires the enzyme lactase to break it down into the simple sugars galactose and glucose. Absence or deficiency of this enzyme makes for upset stomachs with too much dairy consumption. But it's not an allergy. Countless parents swear that their child's colds and/or ear infections subside when they "cut out dairy," often by the recommendation of their doctor. There is no good evidence to prove this.

Finally, I would be remiss not to address the GMO debate, which I'll start here and continue in upcoming chapters. In a classic Jimmy Kimmel bit, he and his team left the studio and went to local farmers' markets in the Los Angeles area and asked "health conscious" shoppers what they thought of GMOs. They unanimously thought they were bad. And, also unanimously, they couldn't recall what the acronym stands for, nor could they begin to explain what could be so bad about them. Genetic modification has been present in some form for thousands of years, beginning with the advent of selective domestic plant and animal breeding as long ago as 12,000 BCE. Purposefully breeding organisms with desired traits is the earliest form of genetic modification. While the genes themselves are not altered, the desired genes are propagated, and the less desired genes are faded out. How should we be thinking about GMOs today given the current science and lack of data? There's definitely reason to be confused, but what about concerned?[19]

Let's first go over what GMOs, in today's parlance, are. GMOs, genetically modified organisms, be they plants, animals, or microorganisms, have had their DNA, or genetic material, altered in ways not seen in natural reproduction. Genetic material is transferred from one organism to another, including between different species.[20] One of the many ways that GMOs get a bad rap is because they may contain toxins. Indeed, they may. One genetic modification on crops has been to incorporate a gene for insecticide production. This is to help plants kill pests, precluding the need for external use of pesticides. The amount of pesticide in the GMO plant is substantially less than the amount of pesticide used on non-GMO plants. But because of this "toxin" in the food, it's portrayed as more harmful than its non-GMO counterpart. Indeed, much needs to be done to better understand the short- and long-term implications of genetic modification, especially when it comes to what we eat. It can be positive—genetic modification to reduce use of pesticides, genetic modification to enhance nutritional benefits, and even genetic modification to reduce allergens and contaminants. But on the flip side, the concerns are real: genetic modification may change the bacterial population of our food, in turn altering our internal gut bacteria, leading to antibiotic resistance.[21]

One of the problems is that commercial food companies fund many of the studies on the healthfulness of particular food ingredients, as well as on the safety of genetic modification. This conflict of interest has led to increasing difficulty in understanding data. For most of my professional life, I have been a proponent of food-safety labeling regarding choking hazards for certain foods. While some foods, such as some brands of hot dogs, have these safety labels, most do not. It's been hard to get food companies who sell these high-risk products to get on board to label their foods for safety. What's in it for them except for lower revenue and, oh, just a few lives saved?

The GMO debate, however, is not quite as clear as life-and-death issues such as risks of allergen exposure or choking hazards. Proponents of GM in foods tout lower need for pesticides, higher yields of crops, and a multitude of studies of large animal and plant populations demonstrating GM's long-term safety. There remains no consensus on GMO safety, and the issue now is mandatory labeling of foods containing GMOs. Because trillions of meals containing GMOs have been consumed, there is no way of prospectively measuring any negative health effects in humans over the short and long term.[22]

Bottom line: Genetically modified foods have been in our diets for the past twenty years. Large studies, as well as large studies of large studies, have found no evidence of harmful impact on our health nor the health of our crops. The science of genetic engineering and modification is expanding exponentially, challenging us even more to be able to identify any harms from these advances. Certainly we can be cautious with new technology, but it should not always be viewed as the enemy: it can minimize the need for pesticides, lead to more nutrient-packed products, and lead to longer-lasting fresh foods, obviating the need for freezing or canning. It sure beats needing to put some powdery potion in your smoothie.[23]

HYPE ALERT

In choosing the "best" diet, remember the six rules:

1. Pick and choose what works for you (there is no "best" diet).

2. Don't be fooled by superfoods; they don't exist.

3. Appreciate your body's own detox centers; "detox" diets can be dangerous in the long term and bank a lot on the placebo effect.

4. Don't be terrified of gluten if you don't have celiac.

5. Keep a food journal.

6. Understand true allergies.

FAT-FREE SUGAR, ORGANIC COOKIES, AND "FRESH" PRODUCE: A WALK THROUGH THE SUPERMARKET

How to Read a Label

> WHY DO PEOPLE WHO EAT GLUTEN-FREE HAVE HIGHER LEVELS OF ARSENIC?
>
> WHAT DOES *FARM FRESH* REALLY MEAN?
>
> WHICH TYPES OF FISH SHOULD YOU CONSUME?
>
> IS THERE A CANCER-BUSTING DIET?

If I asked you to precisely define the difference between *100% All Natural* and *Organic* on your food labels, could you do it? Do you know the rules for calling something *wholesome* or *low-fat*? What about a product that says *Good source of calcium* or *High in vitamin D*? What do these terms mean? Welcome to the marketing hype in our modern food supply. Every time you step into a market, you're encountering vague claims that will dupe you. The term *organic* is one of the worst misnomers of our era. Aflatoxin, which is derived from fungi and found on peanuts, is technically organic, yet it's among the deadliest materials on the planet. Anthrax is organic, too. And arsenic is "all natural" *and* organic. (If you're gluten-free, you may want to have your arsenic levels tested. A 2017 study led by an epidemiologist at the University of Illinois at Chicago showed that elevated levels of arsenic are often found in people who avoid gluten.[1] Why? Rice flour and other rice products that naturally contain arsenic are often used as substitutes for gluten-containing ingredients in foods. So when you eat a lot of food manufactured to be gluten-free, you could be consuming a lot of arsenic.)

Perception is so much of what goes in to making a seductive food label. As good, conscientious humans, we are drawn to words such as *natural, sustainable, fair employment practices, organic,* and *wholesome.* As we should be. But try to look at the other side, because there most often is one. Something that seems so obviously good and right may not be either. A few years back, a friend of mine took a debate course, and one assignment was to debate the issue of recycling. He was tasked to debate against it. "You'll lose," I said. "There is nothing bad about recycling; it's all good. It's the right thing to do." But he had already put some thought into it: "What about the emissions belched out by the recycling trucks? The fumes emitted by the recycling plants, and the wasted work time by all the employees who have to sort through the garbage placed in recycling bins that isn't recyclable? Do you think everyone scrapes their take-out cartons clean before dumping them into the recycler? That just adds more unnecessary waste." Touché. He also won the debate in class.

While recycling *is* a critical addition for our earth's longevity, there's always another side to what seems unquestionably good. So much extra work, resources, and energy can be put into foods, either to add goodness or remove badness, that, as with recycling, there can be costs at worst, or no benefits at best, to supposed nutritional foods. Perhaps even more noteworthy than these unnecessary efforts to remove fat, add vitamins, avoid additives, or free food of harm are the food labels stating something is not there, although it wouldn't have been there in the first place. Gluten-free lollipops, sugar-free coffee, and fat-free juice, to name a few. Smoke and mirrors at their finest.

The truth is that there is almost no regulation, by any group or government agency, regarding most food labeling. And the labels that are regulated mean remarkably little for nutritional value. This applies not only to the marketing angles on packaged foods, but also to the supposed fresh foods, be they *farm fresh, fresh caught,* or *freshness guaranteed.*

A few summers ago, I traveled with my family to Greece. We stayed for a few nights at a small, family-run hotel on a farm. The hotel served vegetables from their garden, eggs from their chickens, and fruit from their trees. The food was fresh. Farm fresh. The yogurt was, you guessed it, Greek. But there they just called all of it food. The people of Greece walk a lot because that's how they get around. They eat when hungry and stop eating when full. The food didn't have labels—it just tasted good and was healthy. For the most part, the Greeks looked healthy, too—not a lot of perfectly toned Adonises, but not many obese cherubs either.

Now let's take a walk through your local supermarket and see what we have. By the end of this tour, I will have saved you a lot of money.

FRESH FACED

Something that many nutritionists recommend is to start your shopping at the periphery aisles—those containing produce, dairy, meats, eggs, and fish. Many items here don't come with a nutrition label. The central aisles, on the other hand, are filled with lots of packaged, processed foods that have long-winded nutrition labels and are more likely to contain artificial ingredients, many of which a biochemist would struggle to pronounce. So let's start with the healthy stuff—good old fruits and vegetables.

Most markets today sell both organic and "conventionally farmed" produce. How to choose? Most health-conscious shoppers will go for the organic items, even though the price is substantially higher. But ask most of those shoppers what they are really paying for, and many of them will not know the precise definition of *organic*.

The organic food industry is a multibillion-dollar one. Rigorous government guidelines must be met to label any food organic, and it all comes down to the farm the food is from. The USDA regulates whether any food can be labeled organic, but of course nothing is that simple, especially a label. *Organic,* whether referring to produce, meat, eggs, milk, or otherwise, means that the farmer did not use pesticides or synthetic fertilizers, and that animals were not given antibiotics or hormones.

Products labeled organic must be at least 95 percent organic; products labeled 100 percent organic must contain, you guessed it, 100 percent organic materials. But look carefully: many labels will state *made with organic ingredients*. The USDA regulates that at least 70 percent of those ingredients must be organic, and no ingredients can contain the dreaded genetically modified organisms (GMOs, see page 98). But the issue is, what does this mean for consumption, as well as for ecological farm sustainability? We know that most organic foods are more expensive than their conventional counterparts, but is organic food healthier? For some products, yes; but for most products, no.

Organic products do not contain more nutritional value than non-organic ones.[2] A 2012 study appearing in the *Annals of Internal Medicine* reviewed hundreds of other studies looking at whether organic foods are healthier than their conventionally grown counterparts.[3] For the most part, the hundreds of studies reviewed by these investigators found no difference in health outcomes in people eating organic versus conventionally grown foods. Yes, more pesticides were on the conventional products, but even these levels were insignificant to the health of the consumer. Organic diets in pregnant women did not lead to fewer food-related allergies in their children, and the vitamin contents of the two types of products had few differences. For many foods, such as fruits and vegetables that have peels (e.g., bananas, citrus, and legumes, to name a few), pesticide use or type of fertilizer makes absolutely no difference. Fruits and vegetables with thick peels, such as mangoes, pineapples, avocados, melons, and eggplants, have virtually no pesticide residue on the edible portions, so save your money and buy conventionally grown ones. Surprisingly, even some without thick skins, such as asparagus, mushrooms, cauliflower, and onions, also have little to no pesticides, making them absolutely safe, even if not organic.[4] When quantifying nutritional outcomes based on eating conventionally versus organically grown foods, it's virtually impossible to tease out the differences. There are just too many variables, such as total consumption, activity levels, and, perhaps most important, basal metabolism of individuals. Not everyone digests the same food at the same rate; not everyone absorbs nutrients identically, even if identical foods are eaten.

Craig Underwood, owner and operator of Underwood Family Farms in Moorpark, California, has been all too aware of the buzzwords thrown around regarding health and safety, feeding into consumers' fears about "conventionally grown" produce. He clearly states that while the term *organic* means pesticide-free, many pesticides are safe; in fact, much safer than the pests that they get rid of. The vast majority of sprayed produce receives infinitesimal amounts of pesticide, similar to a drop in the ocean. "Apples have seeds with cyanide, and pears contain formaldehyde, but we eat them and live to talk about it." Having been in the farming business for decades, he has seen a shift in the term *buying organic* to denote a self-righteousness that is, for the most part, unfounded. While regulations on fertilizer use and pesticides are critical, the negative connotation of anything external, or not "natural," has become a marketing tool nonpareil. He loves to mention that there is now organically grown tobacco, despite that nicotine is a well-known, and effective, pesticide, let alone a toxin when inhaled or chewed. And just as one can drink "gluten-free" coffee, the term *organic* is overused on foods that are so hearty that no pesticides would ever be needed in the first place. "Folks should stop worrying about the produce section in their supermarkets. It's the stuff in the center aisles [meaning chips, sodas, and baked goods] that could kill them." Wise words, Mr. Underwood.

Next question: What about the label *farm fresh*? What kind of farm, and how fresh is fresh? Contrary to how one might envision *farm fresh* to mean, fruits and vegetables can be labeled fresh as long as they've not been cooked or frozen.[5] But even if they're not packaged for the frozen-food aisle, they can be stored on ice or in containers and still sold as fresh. Fresh produce at farmers' markets is exempt from the latest food-safety regulations, so what does that mean in terms of health risks? If grown on an organic farm, the produce may not contain pesticides, but it can be covered with wax, which isn't terribly dangerous, but who wants to eat fruit covered in wax? It most definitely did not come fresh off the tree that way. And what's truly better—fresh, canned, or frozen? Many frozen fruits, especially berries, contain the same vitamin and fiber content as their fresh counterparts (or even more so, because they are picked at the peak of ripeness and flash-frozen, as opposed to their comrades that are picked before fully ripened in order to survive the long journey to your fresh-produce section). This is critical, as frozen berries and other healthy fruits can be bought and stored all year round and are significantly more affordable than the fresh ones, especially in the off-season. Many consumers will ignore the frozen-fruit or frozen-vegetable aisle, purchasing only the so-called fresh stuff. But these frozen goods can be more nutritionally dense, as well as more affordable and last longer. Canned fruits and vegetables, on the other hand, are less ideal. While some do contain similar amounts of vitamins and fiber, many will lose the benefits of fruits and vegetables (which are primarily vitamins and fiber) during canning, when they are cooked to nearly gelatinous consistency. Even worse are the fruits canned in that sweet syrup, adding unnecessary sugar to the mix.

FISHY BUSINESS

Regardless of your dietary restrictions—vegan, pescatarian, kosher, or otherwise—let's head to the meat section, where we choose between red meat, fish, and fowl. Some are captured wild, and some are raised on farms.[6] The question becomes, Do these farms use traditional practices or do these products come with a *free-range* label? What were those animals fed? Did they receive hormones and/or antibiotics? And, most important, does any of this matter?[7]

Let me state one fact from the get-go: we don't eat enough fish in this country. Many people avoid it simply because they don't like it. But it is high in quality protein, low in fat, and high in the famed omega-3 fatty acids, which can reduce heart disease and stroke. While wild-caught fish have been found to have increasingly more chemicals such as mercury and PCBs (a group of man-made chemicals banned in 1979 that last a long time in the environment and build up in fish) from our polluted oceans, rivers, and lakes, and some are overhunted and are becoming endangered, certain fish still provide benefits. The worst offenders are the fish that live long and are higher up on the food chain. These giants of the sea—such as marlin, tuna, shark, swordfish, king mackerel, halibut, bluefin tuna, albacore, tilefish, and northern pike—bioaccumulate the toxins from all the fish below them in the chain. The swordfish that eats the tuna that eats the salmon that eats the anchovies. The fish least likely to be as contaminated are the smaller, shorter-lived ones: salmon, pollack, anchovies, sardines, herring, sablefish/black cod, and sole. Those are the fish to seek when you plan your meals.

The aquaculture industry has surged, with many fish now being "farm raised," in efforts to create healthier entrées. But (there's always a but) these farms may also be fraught with problems—altering the surrounding ecosystem in their body of water, polluting the surrounding seas with organic waste (fish poop), and altering the natural species populations surrounding these farms. The jury is still out regarding the choice of wild or farm-raised fish. Certain species are known to have higher mercury content than others, and their intake should be limited by pregnant women and young children, but the overall concern with fish is that it's not consumed enough. Fish will be labeled *farm raised, wild caught, fresh frozen,* and *no added color.* Just get some fish, and don't worry too much about the labels. If you have a choice between, say, salmon and swordfish, go with the littler guys (salmon).

When we think of GMOs, nuked fruits and vegetables often come to mind. But the ocean waters are now getting filled with megafish. As genetic modification can occur on all levels, including with humans, fish have not been out of the loop. The company AquaBounty Technologies has engineered a breed of salmon to grow to be five times the size of their nongenetically modified counterparts. Since the gene added is found in an existing breed of fish, the GMO label is not required by the FDA. Concerns about sending these farmed fish out into the oceans, altering the natural ecosystem, are hotly debated.[8] For now, I'd stick with fish caught in the sea—not bred in a factory aquarium.

SURE AS EGGS

The egg section, which now can take up a large space in a refrigerated area, is great. Your choices used to be straightforward: white, brown, large, or small. And grade—that perennial USDA stamp that somehow gives an egg its worth. Grading an egg has purely to do with how the shell looks, thereby indicating only that the yolk and the white have a good appearance.[9] The stamp has nothing to do with safety, absence of contamination, or quality of the chickens. Have you ever bought a Grade B egg? I've never seen a Grade B label, but they do exist and are barely inferior to their Grade A neighbors. Those poor, above-average Grade B chicken embryos! They'll never get accepted at a top egg college with B's. But Lake Wobegon grades are no longer a focus, as we now also have choices about how those chickens that laid the eggs were treated before donating their unborn babes to us: farm-fresh eggs, eggs produced by free-range chickens, eggs produced by chickens that ate the organic food you just bought, and all-natural eggs.

Many of us who live in cities or the suburbs, despite our access all day and night to news of real-life farms, still prefer a bucolic image of farm life—early mornings, roosters crowing, chickens clucking, and biscuits baking. The stereotypical farmer (always a man) is out early with his straw hat, greeting each animal by name as he sprinkles some feed into the chicken coop, throws the dinner leftovers into the pigsty, and hands a clump of grass to his favorite cow. The farmer's wife, apron on, is always with a big, happy belly, and rosy cheeks, baking away and tucking her children under her skirt as they scurry around the house. Wipe that out! Not farming these days.

Farming is an incredibly important part of our planet's food supply, but it's not Farmer John and Wife Jane. And the "farms" where farm-fresh eggs are produced and free-range chickens are roaming are anything but fresh or free. Farm-raised chickens live in cages, with many cages in one space. Until recently, each chicken would get about 67 square inches of space. That's about 8 by 8 inches, the size of your smallest baking pan. Some farms have now increased that space to 116 square inches—and to save you the calculation, that's about 10½ by 10½ inches per chicken. *Farm fresh* means nothing. Even the *cage-free* ones live in large, overcrowded barns, filled with thousands and thousands of chickens. Each gets about one square foot of space. The *free-range* ones are lucky enough to get a few minutes each day to stretch their little legs, but even these are kept in incredibly tight quarters. Because they live in such tight quarters, without the cage barriers, they are twice as likely to die, by either illness or pecking each other to death, than their caged fine-feathered friends. Those ranges, by the way, are smaller than this book. So these cramped free-range, farm-fed, cage-free, fresh chickens beget your free-range, cage-free, farm-fresh eggs. And they're organic![10]

When it comes to labels on eggs such as *hormone-free* and *antibiotic-free,* don't be fooled. It is against the law to inject hormones into chickens, although they do receive antibiotics along with their bovine neighbors. Even if they do receive hormones, they do not remain active after they've been digested (and if they were active, then diabetics would not have to inject themselves with insulin—they'd be able to just take a pill!). Antibiotics, on the other hand, continue to wreak havoc on the food industry—not just in chickens, but in most farm-raised meat we consume. Just as giving antibiotics to humans unnecessarily leads to antibiotic resistance and more aggressive infections, unable to be treated with antibiotics, the same goes for our avian friends. When a chicken egg is injected (yes, eggs are injected with antibiotics, one by one even before the chickens are hatched), the stronger microbes survive, leading to widespread antibiotic resistance in poultry.[11] Slapping claims on the labels will make your eggs just as healthful as your gluten-free coffee. And what should these (not so) frolicking chickens eat? Many egg boxes, or chicken products, will claim their chickens are vegetarian. Well, this is not so good. Chickens are omnivores and subsist on worms as well as grains in the wild. The vegetable matter they're given is likely a corn feed, pumped with amino acids.

GOT ORGANIC MILK?

On those farms, the cows need to be milked! There are free-range cows, organically-fed cows, cows given hormones, and cows given antibiotics. Cows allowed to roam, and cows locked up. Cows that produce organic milk. Cows that don't need their milk pasteurized. And then there's the fatness—fat-free/skim, nonfat, low-fat, reduced-fat, regular. Purchases of organic milk, especially by households with young kids, are a large reason why the Organic Trade Association leaped from a $3-billion-per-year industry to one north of $30 billion. Increasing concerns about additives in milk, such as growth hormone and sex hormones, and pesticides in the cattle feed and antibiotics administered to cows, have led to more and more consumer concern about milk not labeled organic. The vitamin, protein, mineral, and fat contents in milk are the same in organic and conventionally processed milk.

I never gave much thought to my brand of milk until I had kids. Growing up, we drank what my friends would call blue milk (skim had a bluish hue, I suppose). I continued with this habit into adulthood. The rare instances I came across whole milk, I felt I was drinking ice cream. But when my own kids were old enough to begin drinking cow's milk, I, too, fell into the mass milk-mania megamarket. I bought whole organic milk, preferably labeled with as many *withs* and *withouts* as possible: With added DHA! Made from cows without exposure to rBGH! With vitamin D and calcium! And my favorite, truly showing how much of a sucker I was: Our cows are happy cows! Yes, I still do buy organic milk, but mainly out of habit. The regular stuff is certainly fine by me (and my family) as well. I'm not so sure if it's from particularly happy cows.

The big bad one is raw milk, meaning unpasteurized. This pseudo-healthy form is not organic—it simply does not undergo pasteurization, a method of removing harmful bacteria, first developed by Louis Pasteur in the late nineteenth century. Raw milk has known pathogens, such as *Salmonella, E. coli,* and *Listeria,* to name a few, and has been responsible for countless outbreaks of human illness. It is not recommended by either the American Academy of Pediatrics nor the Centers for Disease Control and Prevention nor the FDA.[12]

After all is said and done, the milk has additives—DHA is a big marketing tool. As with most acronyms, most who swear by it neither know what it stands for nor know how it's supposed to work. And they surely don't know the direct implications of its being added to milk, or any other product. DHA, or docosahexaenoic acid, is a highly unsaturated omega-3 fatty acid (HUFA). It is found naturally in marine algae, and thus in certain types of seafood. It is also present in human milk and is important for brain and eye development in young infants.[13] Because it has been linked to neuronal development, a surge of studies have evaluated if it may impact neurologic disorders, especially in children. DHA is added to store-bought milk, or taken in supplements, in an attempt to stave off developmental disorders, including ADHD and autism spectrum disorders. Some studies have demonstrated potential benefits of omega-3 fatty acids in treating mood disorders in adults.[14] However, the organic-food-industry watchdogs were up in arms when word got out that the added DHA in organic products, including milk, was synthetically processed DHA.[15] Although DHA is DHA, no matter how it's processed, many went batty at the thought that something synthetic was being added to foods labeled organic. Although the FDA requires the organic label to mean that a product contains only organic ingredients, the agency let the DHA stay.[16]

Moreover, these DHA-infused products include GMOs.[17] Industry watchdogs also claim that the synthetic DHA contains harmful substances called hexanes, thought to be neurotoxic, causing more harm than good to our nervous system. Here's the bottom line: DHA is a critical substance for adequate eye (specifically the retina) and brain development in the first six months of life. Human breast milk contains it, and infant formula should have it if a child is receiving formula in place of or in addition to breast milk. But after that, added DHA, be it GMO, synthetic, or organic, probably makes no difference.[18] More recently, evidence is developing that added DHA makes absolutely no difference in an infant or child's development. It won't make them smarter, either.[19]

Addressing the real or perceived benefits of DHA, organic, or any of the other myriad claims such as *no rBGH* (recombinant bovine growth hormone); *lactose-free; produced without the use of pesticides, antibiotics, or hormones; our cows make milk the natural way;* or *a clean-living cow . . . makes really good milk* is enough to drive a consumer to drink. But these labels, with or without using the term *organic,* do work in marketing. People will pay a little more for a product claiming to be healthier. There's even a term for this: the *hedonic price function.* When it comes to perceived health benefits (or absence of health harms), hedonism rules. Much to my surprise, and likely yours, the American Academy of Pediatrics has found that most of these labeled items make little or no difference, including rBGH, which affects the cows, but not their milk. Nor do antibiotics, pesticides, or organic cow food.[20]

Claims for dairy's benefits include its vitamins, minerals, and protein. Dairy products have significant amounts of vitamin D, calcium, and protein. But how do labels claim *source of, good source of, excellent source of,* or *high in . . .* ? Other dairy products, such as yogurt, can squeeze in some more goodness by the labels *active cultures, probiotics,* and *Greek.* (Greek yogurt is simply regular yogurt with much lower water content. This creates a denser flavor and taste, with relatively more protein per serving. In Greece, it's called yogurt.)

As dairy does have so much goodness, let's see what all these terms really mean. Most cows that produce milk do so in factories, even if they are called farms. It's a factory on a farm, perhaps, but it's a factory. Wipe away that image of the farmer sitting on a milking stool, squeezing each teat, hoping not to get the occasional kick from Elsie or milk squirted in his eye. Milking farms are large-scale, automated, machine-run factories, where cows stand in close quarters for much of the day, as the milk is expressed by machines. Not into buckets, but into other machines, where the processing begins. That's *farm-fresh* milk. Vitamins, minerals, and protein occur naturally in milk, as does a small percentage of fat (full-fat milk is 4 percent, low-fat is 1 or 2 percent, and nonfat is 0 percent). The multiple types of milk have surprisingly little difference in fat content, but the lower the fat content, the lower the protein content and the higher the relative sugar content. So you may be cutting a few (very few) calories drinking the low-fat versions, but you'll also unknowingly cut down on protein and add some sugar to your glass. Milk also contains the naturally occurring sugar lactose. Many folks claim a milk and/or dairy allergy when it is a lactose intolerance or sensitivity. Lactose is a disaccharide sugar, meaning it is a double, or complex, sugar. Our digestive systems contain the enzyme lactase, which breaks down lactose into the simple sugars glucose and galactose. Lactose-free milk comes with this process already done. Glucose and galactose taste slightly sweeter than lactose, which is why lactose-free milk might seem sweeter than the regular kind.

Vitamin D and calcium occur naturally in milk, but in substantial amounts? Yes. I've always been a big proponent of milk and dairy products, which have all gotten such a bad rap over the past decade or so. Parents swear that their child's ear infections subsided after they stopped dairy. No! Please don't think that! If a child has a bottle or sippy cup of milk while lying in bed, right before falling asleep, then, yes, that kind of dairy consumption can lead to cavities, nasal congestion, and even ear infections. If you lie flat while drinking, especially infants or toddlers, the milk tends to sit at the back of the throat or the top of the esophagus. It can then easily head behind the throat, toward the back of the nose. This is where the ears drain, and thick milk at the back of the nose can lead to poor ear drainage, nasal congestion, and ear infections. The sugar from the milk in either a bottle or a sippy cup can cause what's called bottle rot, which is seen in older babies and toddlers who have a bottle in their mouth for hours on end. The damage to the teeth can be done even at the level of the gums, so babies without teeth are at risk for cavities in their future from taking the bottle to bed. But dairy itself is not associated with ear infections, nasal congestion, or cavities. People also consider dairy to be a trigger or extender of colds. So many patients of mine have been under the impression that eating such treats as ice cream or any other milk product after tonsil surgery is out of the question. Why? There is absolutely no evidence that it slows the recovery or causes any problems. A child with a sore throat given ice cream is a happy child.

IN THE MIDDLE

The central aisles of the supermarket are perhaps the most label laden. Almost every product claims to either have more of the good stuff or less of the bad stuff. Or both. *Good source of calcium! Low in saturated fat! High in fiber! Enriched with whole-grain goodness!* Beyond *excellent source of,* we have *made with, contains,* and *packed with,* in efforts to boost seeming healthiness of products and sway the consumer to buy. The other angles are *low in sugar, low-fat, whole-grain,* and *all-natural.* Many of these terms don't have much regulation. For instance, for a product to claim *good source of* (something good, no doubt), it need only have 10–19 percent DV (daily value) of the RACC (reference amount customarily consumed). This is minuscule, but it sure does sound good. And you can be sure that if *good source of* claims and the like require a minimum of 10 percent, the makers will go for the minimum, not the maximum.[21]

The designations *made with* or *a touch of* mean just as close to nothing as you can get. I love Tazo organic green tea. The four words right there pack a punch: *Tazo,* a healthy brand, *organic, green,* and *tea.* Health in a bottle! It gets better: *with just a touch of sweetness.* Yeah, thirty grams of sugar per serving. Pretty heavy touch. I love the honest labels: *contains 10 percent real fruit juice!* (Exclamation point included!) Fill an eight-ounce glass with water. Add three tablespoons of sugar. Then add about half a shot glass worth of juice—there's your 10 percent. And while we're at it, the "press" that juice gets is much too good for its own worth. Even better are the brutally honest ones: *contains no real fruit.* Now we're talking.

The vast majority of juices have little or no nutritional value—even the 100 percent ones—the value was in the fruit itself, which was rid of fiber, its most valuable commodity, to make it into juice. Some juices, such as apple or orange, have some vitamin C, but for the most part the high sugar content of even *100 percent fruit juice* far outweighs any vitamin benefit. When my kids were younger, they knew that juice was something they'd get at birthday parties, and they knew that I called it sugar water, which is what it is. They would always be polite and say to the host, "Thanks for the sugar water!" The host would look puzzled, then get it, and pass me a sneer with my slice of cake.

Sugar has increasingly been called a villain over the years, and in many ways this is well deserved. It became a more and more prominent ingredient in the low-fat foods, with the erroneous notion that low-fat must equal low calorie must equal healthy. In the 1970s, the low-fat generation led to foods with higher and higher sugar content. This led to more obesity, not less. More type 2 diabetes, not less, and even more heart and liver disease.

However, even blameworthy sugar isn't as bad as all that. While it may lead to a long list of preventable, chronic illnesses, it isn't the evil substance that some think gets sprinkled on cells to feed cancer. A small but growing group of cancer doctors, and, as a result, cancer patients, subscribe to the notion that because cancer cells need sugar to grow, a no-sugar diet can treat or even cure cancer. Low-sugar diets and anti-cancer "detox" naturopathic centers like to make money from this claim. This notion is a complete misunderstanding of something called the Warburg effect. Warburg was a scientist who demonstrated that cells, especially rapidly growing cancer cells, need sugar for energy and cellular multiplication. However, this does not directly correlate with sugar in one's diet. Just because cells in a petri dish need sugar to grow has no bearing on cancer growth in humans. Plants give us oxygen. But if you are short of breath, would you go smell a flower? However, plenty of shady doctors, and desperate patients, subscribe to this sugar fallacy.

A patient I know who had a curable tongue cancer chose no-sugar dieting over surgery. The cancer grew rapidly, invading vital structures, and the patient was no longer a candidate for surgery. As he became well aware then that the sugar-free diet was not an option, he agreed to chemotherapy and radiation treatments. But the cancer had grown so rapidly during his diet of no treatment that it did not respond to these salvage therapies. He eventually lost the ability to eat or talk and later was unable to breathe as the tumor blocked his airway. He underwent palliative placement of a tracheotomy tube and died weeks later in a hospice.

Dietary sugar is also falsely connected to yeast infections. Not so. I love how one doctor puts it: "The sugar-consumption/yeast connection is an urban myth, perpetuated it seems both by many well-meaning, but ill-informed, health-care professionals as well as purveyors of snake oil (you know, the ones who want to sell you the cleanses, diets, and books designed to help you rid your body of yeast)."[22]

In finding healthy foods in the hopes of being healthy, fitness trumps fatness. In other words, the obsession in our culture to be thin can be deadly. So many well-meaning dieters have missed the boat on this. Healthy fats are part of a healthy diet. Some low-fat or nonfat products lose so much on the nutrition side, substituting sugar for fat, or fake fat or fake sugar, that the calories "saved" are hardly worth it. Dr. Carl Lavie, author of *The Obesity Paradox,* demonstrated that fat-free people are not necessarily healthier, in the short or long run.[23] And you can look thin but still have too much visceral fat—unhealthy fat deep in your midsection around your vital organs that increases your risk for metabolic disorders and even overall mortality. So you probably don't have to lose those last ten pounds, especially if you're north of age fifty and at higher risk for all kinds of illnesses largely related to growing older. Unfortunately, those who focus on omitting specific elements of food (namely, for dieting purposes, sugar and fat) make up the calories with unhealthy choices. New research is also showing that it's more important to focus on staying active than to harp on diet alone. Constant movement may do more for your health—and metabolism—than any special diet can.

HYPE ALERT

* Most food labeling is unregulated by any group or government agency. The labels that are regulated mean remarkably little about nutritional value.

* *Organic* does not equate with *healthy*; for some products, yes, but for many, no.

* Eat fish caught in the sea, not bred in a factory aquarium.

* Dairy is not associated with ear infections, nasal congestion, cavities, or longer colds.

* You cannot prevent (or cure) cancer through diet alone.

* When reading labels, know your terms: *high in, good source of, no added sugar,* etc.

7

THE TRUE COST OF BEING FORTIFIED: SUPPLEMENTS, POWDERS, AND POTIONS

How to Remain Vital Without Vitamins

> WHY DO WE LOVE OUR VITAMINS SO MUCH?
>
> WHICH VITAMINS CAN INCREASE RISK FOR CANCER?
>
> WHAT HAPPENS IF A NEWBORN MISSES OUT ON A VITAMIN K SHOT?
>
> SHOULD SUPPLEMENTS BE CONSIDERED MEDICATIONS?

We love our vitamins. Half of all American adults take a multivitamin or another vitamin or mineral supplement regularly. And 70 percent of people who are sixty-five or older take them. That amounts to more than $12 billion per year. According to Johns Hopkins, that money might be better spent on real sources of vitamins such as fruits, vegetables, healthy carbohydrates, and dairy products. In a 2013 editorial in the journal *Annals of Internal Medicine* titled "Enough Is Enough: Stop Wasting Money on Vitamin and Mineral Supplements," Johns Hopkins researchers reviewed the data, homing in on three particularly alarming studies.[1] One was an analysis of research involving 450,000 people, finding that multivitamins did not lower risk for cancer or heart disease. Another tracked multivitamin use of 5,947 men and their mental functioning over twelve years and found that multivitamins did not reduce risk for mental declines such as slowed thinking or memory loss. The third was a study of 1,708 heart attack survivors who were randomly split into two groups. For up to five years, one group took a high-dose multivitamin and minerals and the other took a placebo (the patients, the physicians caring for them, and study personnel did not know who was receiving a placebo or multivitamin-and-mineral pills). Rates of heart attacks, heart surgeries, and deaths later on were similar in the two groups.

What's more, two other prominent studies spanning more than a decade, completed in 1995, and following tens of thousands of people showed that beta-carotene and vitamin E supplementation in particular can be downright harmful, significantly increasing the risk of cancer.[2] One of these entailed the landmark Framingham Heart Study, which had to end early because the investigators found that taking too much beta-carotene, at first thought to be a cancer-preventing supplement, was associated with significant *increase* in cancer-related deaths. All of these studies have been routinely reported in the media. But they don't seem to change people's nutritional-supplement habits. They don't even sway me. I take a daily multivitamin. And vitamin C. When I have a cold, I go wild and take two C's. And I even take vitamin D and calcium when I'm in the mood. My husband, also a doctor and a researcher, takes his share, and until our dentist implored me to stop, my kids had their gummy multis every morning.

The chief reason I take vitamins is for their placebo effect. As I explain in Chapter 9, this is the idea that, if one thinks something helps, it does. I feel "healthier" taking vitamins. I feel that if my kids go heavy on the fries one day and skip their green beans, they may have gotten some much-needed nutrients in their vitamin.

I will not, however, megadose on anything (double- or even triple-dosing vitamin C, even up to 1,000 mg, is useless, but if used rarely is quite safe), and I would never rely on vitamin supplements to support genuine nutritional health or well-being. Vitamin boosts in drinks are powdery wastes of time and money. Are they dangerous? Probably not. Is there any evidence anywhere that they improve health? No. I work with nurses and doctors who are highly educated in science and medicine and are also quite health conscious. They exercise regularly and often follow their workouts with high-dose protein shakes and mega-vitamin-boosted juices. Will their muscles get bigger because of these drinks? No. But will these boosted drinks harm them? No. Their muscles get bigger because they lift weights, swim, or run. But again, I fall victim. As a busy surgeon, a protein bar of my brand of the month is often my lunch.

I routinely take completely unscientifically validated, yet entertaining, polls of medical staff—surgeons, nurses, residents, and medical assistants—regarding their vitamin intake. The simple question "Do you take vitamins?" leads to some pretty interesting answers, and no two are the same:

"I don't . . . but I should."

"No. Should I?"

"No, but my mom, who's a pharmacist, ships me calcium from Amazon Prime."

"I stopped taking multivitamins because I ran out of them."

"I take E, C, D, B_{12}, a multi, glucosamine, calcium, and fish oil."

"Not really. Why? Okay—no, I don't."

"Every day."

"No, but my kids do."

The use of vitamins and supplements continues to be endlessly debated, with the debate often extending to the fortification of foods and beverages with vitamins and minerals. Do immune boosters in smoothies, including high-dose vitamin C, have any science behind them? Can you take something before flying to help you avoid picking up a cold in the dirty, recirculated air of a plane? Is saw palmetto effective for prostate health? Part of answering these questions entails a discussion about the psychology of taking vitamins and supplements. I'm a prime example: I know all the science that says they aren't beneficial and yet I still take them! Why? And why does the average person trust vitamin companies more than FDA-approved pharmaceuticals? What's the difference?

Because of the minimal regulation of the contents of supplements, be they traditional vitamins or homeopathic herbs, one can never be sure of the actual contents. In 2012, the Department of Health and

Human Services recommended that the FDA have more regulatory input on contents of supplements. Despite this, the FDA has no control. A 2015 investigation of several large retail stores, including Walgreens and Target, found that herbal supplements such as ginkgo biloba and saw palmetto contained anything but these ingredients. Around the same time, the company ConsumerLabs.com analyzed the actual vitamin content of several major vitamin supplements and found drastic differences in the amount of each vitamin compared to what the labels claimed. So aside from the quandary over whether to take vitamins, minerals, or homeopathic herbal supplements, we must question, Are we even getting what we think we are?

A SEMI-VITAL STORY: FROM LIMEYS TO LINUS

Vitamins are essential for life and, by definition, are organic compounds required by a living organism. Currently there are thirteen of them. The two main subsets of vitamins are fat soluble (can be dissolved in fat) and water soluble (can be dissolved in water). The fat-soluble vitamins—A, D, E, and K—can be stored in the liver and fatty tissues until they are needed, which is why we can get by without consuming them regularly. The water-soluble vitamins, B (which has several forms) and C, cannot be stored, which is why we need to consume them more regularly, and why "megadosing" or overdosing on them carries less danger. Early life-forms, several billion years ago, were able to make their own vitamins, but most species, ours included, evolved and lost that ability.

This loss of vitamin-manufacturing ability likely stems from gene mutations over hundreds of generations. Early primates lived in tropical regions containing vitamin-filled fruits and vegetables. Access to these in the diet of our ancestors likely led to the loss of certain gene activity that had been responsible for vitamin production in earlier life-forms. While vitamins are made by living cells, humans can only synthesize two of the necessary thirteen—vitamin D is made in our skin, as sunlight enables the breakdown of a form of cholesterol, and vitamin K is

produced in our livers. Other than those two, we need to take in vitamins from other sources, including bacteria, fungi, and plants. One
hundred million years ago, our primate ancestors made their own vitamin C, but then because of mutations in our human code we lost
the ability to carry out all the steps in synthesizing vitamin C.[3] Now
we rely on plants for this vitamin, as did the hunter-gatherers tens of
thousands of years ago.[4]

Some may say that the term *limey* is pejorative when referring to
an Englishman. But they earned this title for good reason. Despite its
connotations in modern times, the term is a compliment. In the seventeenth and eighteenth centuries, as well as in many prior, likely
including when Columbus was headed west, explorers and sailors at sea
for weeks or months on end would often become deathly ill from an
unknown cause—they would develop full-body aches, purple spots
on their skin, swollen gums, rotting teeth, and sudden death. Unbeknownst to them, they were suffering from scurvy, a severe form of
vitamin C deficiency. British sailors in the late 1800s realized that the
lack of fruits and vegetables on these long voyages led to this disease
and would travel with limes and lemon juice to avoid the illness. It
worked (though at the time no one knew why). Thus the nickname
limey. Not until the twentieth century did the ingredient that prevented,
and also cured, scurvy become known: vitamin C. In 1928, Albert
Szent-Györgyi, a Hungarian biochemist, was one of several scientists
to discover that the body requires certain organic substances, known
as vital amines, in small amounts, the absence or deficiency of which
would lead to various diseases such as scurvy.[5]

Also in the 1800s, manufacturers began to process rice in mills,
which stripped the outer layer of the rice grain. Soon after this, and as
the consumption of rice became more widespread, a disease began in
which people began losing feeling in their legs and were unable to walk.
A few decades later, chickens developed a similar disorder. Those chickens
had been fed primarily white rice. Once chickens were fed unprocessed
(cheaper) rice, they remained healthy. In 1880, the Dutch physician
Christiaan Eijkman found that both humans and avians eating processed rice were suffering from the ailment beriberi due to thiamine,

or vitamin B_1, deficiency. The outer layer of the rice grain is high in vitamin B_1, and stripping this away led to vitamin deficiency.

Vitamin deficiencies are usually reviewed in biology classes, and even in medical school, in a historical context. In my daughter's middle school science class, the students learned about rickets, a bone disorder due to vitamin D deficiency. She learned that this deficiency was not due to lack of vitamin D pills, but due to lack of sunlight in the eighteenth century for children working in a factory all day, then spending minimal time outdoors in fume-filled, sun-obscuring air. Many of these deficiencies still exist in developing nations, as well as in pockets of poverty in the United States. But for the most part our overfed society is anything but vitamin deficient. So why are we under the impression that vitamin supplements, or "boosts," make us healthier? The overwhelming majority of people get substantially more vitamins than are needed for sustenance and growth from a standard (even unbalanced) diet. So why are over half of us hooked on our daily pills?

Just as names such as Andrew Wakefield and Jenny McCarthy are linked to igniting the vaccine controversy, Linus Pauling is the name we associate with the vitamin hoax that lives on today. In 1931, when American-bred Pauling was just thirty years old, he published a paper in *The Journal of the American Chemical Society,* providing an esoteric explanation of how chemical bonds are more complex than previously thought.[6] This led to his winning the Nobel Prize in Chemistry in 1954. Subsequent to this landmark study, he published several historically significant articles about protein structure, evolutionary biology, and molecular biology. He was also a world-renowned pacifist, who opposed Japanese internment camps in the United States during World War II. He derided Robert Oppenheimer's work on the Manhattan Project and declined an offer to work for it, and he vocally opposed nuclear arms. This led to his second Nobel—the Peace Prize—in 1962. He racked up multiple honorary degrees and awards over his lifetime, including the National Medal of Science and being named by *Time* magazine in 1961 one of fifteen scientists as "Men of the Year." He was a hero on so many fronts—a scientist, a humanitarian, a peace activist.[7] But then things got strange.

In 1965, he attended a talk from a so-called doctor, Irwin Stone, a recipient of an honorary Ph.D. from the Los Angeles College of Chiropractics. Dr. Stone offered that the secret to longevity and feeling great while living a long life was vitamin C. He suggested that Pauling, then living the good life in Southern California, take a whopping 3,000 mg per day of vitamin C to feel better and live longer. Maybe it was the California sunshine that is so known to fry one's brain that fried Linus Pauling's into believing this. He soon claimed that taking fifty times the recommended daily dose of vitamin C (yes, that's right, adults only need 60 mg/day, even though today's vitamin C supplements range from 250 mg to 1,000 mg) made him feel more vibrant. And the "vitamin C fights a cold" myth came from this one quote: " . . . the severe colds I had suffered several times a year all my life no longer occurred. . . . I increased my intake of vitamin C to ten times, then twenty times, and then three hundred times the RDA: now 18,000 milligrams per day."[8] The first bona fide megadoser. And vitamin C—not protein chemistry, the advent of molecular biology, evolutionary biology, nor peace activism—is what Linus Pauling will evermore be remembered for.

Maybe it was his advanced age (although sixty-five is hardly advanced in this day and age). Maybe it was the warm California sun. Maybe it was the positive vibes of his alternative lifestyle, the likes of which always seem to start on the West Coast. Maybe because he had been such a superstar in science at such a young age, he began to believe his own press releases. For whatever reason, Dr. Pauling had clearly gone off the deep end with vitamin C. Soon after he started making his claims in the early 1970s, he wrote an international bestseller at age seventy called *Vitamin C and the Common Cold,* later reprinted as *Vitamin C, the Common Cold, and the Flu.* In both books he recommends at least 3,000 mg per day of vitamin C to ward off these pesky common ills. What followed was the beginning of the vitamin C craze, with over 50 million Americans, me and my family included, taking vitamin C supplements to avoid and treat colds. I remember the slightly chalky, tart orange chewables I took daily as a kid. And I'd take two when I had a cold. Sometimes I didn't feel like taking them, so I snuck them to my dog.

But here's the irony: countless large studies, looking at thousands of subjects, in a multitude of medical centers in the United States, Canada, and abroad, found no benefits from vitamin C supplements—no reduction in the incidence, severity, or duration of colds or the flu. But to this day, vitamin C remains not only one of the most common supplements sold, it's now also sold as boosts in smoothies, or in packets to repel the evil humours while traveling. There is no evidence that these superboosts make any difference in frequency, severity, or duration of colds.

Linus Pauling later claimed that it wasn't just vitamin C that prevented the common cold—it was several vitamins, specifically vitamin C, vitamin E, and beta-carotene (a precursor to vitamin A), as well as selenium (an element in such foods as tuna, Brazil nuts, halibut, turkey, and beef), that could treat cancer as well as multiple other diseases, including heart disease, infections, diabetes, fractures, altitude sickness, depression, and rabies, to name just a few on his exhaustive list. Because his name and prior research carried so much weight, even with his recent extreme statements, several academic centers, including Johns Hopkins, the Cleveland Clinic, the National Cancer Institute, and the University of Minnesota, enrolled hundreds of thousands of patients in vitamin studies, specifically looking at increased or decreased cancer risk. Every study, which included close to 1 million subjects, found an *increased* risk of cancer when taking some of these supplements touted to stave off malignant disease. The Framingham Heart Study, which was initially set to cover many years, had to break the study's double-blinded code early because of the dramatic, substantial clear detriments from beta-carotene supplementation in the cohort of the study population taking high-dose beta-carotene.

FREE YOURSELF FROM RADICAL BUT STILL BE SMART

The theory behind the supposed prevention and even cure of cancer by many supplements had to do with antioxidants. No doubt you've heard about antioxidants. The word still buzzes across airwaves, grocery stores, antiaging beauty bars, and even juice bars. Those special substances are advertised as antagonists to adversarial free radicals floating in one's cells—those "radical" entities thought to be the nasty culprit of cancer development. It is more likely that having some free radicals banging around isn't such a bad thing, and perhaps they shouldn't be bound up by overdosing on antioxidants. A little radical in everyone may actually do some good and attack some of the developing cancer cells. While the precise balance is not yet known, it is known that vitamins do not prevent colds, the flu, cancer, or the blues. Yet we still buy them, take them, and even recommend them.

Despite the hype in taking copious vitamin supplements via capsules, tablets, water, and powders, some vitamins are critical to take. Newborns are given a lifesaving vitamin K shot in the first hours of life. However, this vitamin is delivered via an injection, so the anti-vaxx crowd has expanded its nonsensical crusade against anything-injectable-because-it-is-unnatural-and-unnecessary-and-will-cause-harm-even-cancer to include it. The anti-vaxx community advocated for either no injectable vitamin K to newborns or administration of vitamin K in oral form. Newborns are given vitamin K to prevent hemorrhagic disease, or what's now known as VKDB (vitamin K deficiency bleeding). Without the vitamin K shot, up to one to two newborns out of one hundred will develop VKDB in the first weeks of life, which appears in its mildest form as nosebleeds or intestinal bleeding, or at its more common severe form as sudden bleeding in the brain. Vitamin K acts in the clotting cascade of factors in the blood, and babies are born with inadequate amounts of this vitamin (and breast milk contains just a small amount). Adequate vitamin K stores in the body do not develop until a baby is between three and six months old.

The importance of vitamin K in blood clotting was discovered by a Danish biochemist in the 1930s (the *K* stands for *Koagulation*). He found that chickens who were not given a diet with enough of this substance had higher rates of bleeding and bruising. This led him to evaluate the presence of vitamin K, or *Koagulationsvitamin,* as an essential portion of the ability of the blood to clot. It is exceedingly rare for a child or an adult to have vitamin K deficiency, as such a small amount is needed for the blood to clot. But there's enough concern among doctors to prevent life-threatening bleeding in newborns that the American Academy of Pediatrics issued a statement in 2003 encouraging physicians to give all newborns a vitamin K injection at birth.[9]

In several small studies, later found to be inaccurate and unfounded, injected vitamin K increased the likelihood of childhood leukemia. This and other misguided worries about injections in general has led to an increase in VKDB in this country, with more and more infants succumbing to VKDB, because misinformed parents decline the injection.[10] The story of baby Olive is tragic. At just four weeks old, one evening she was found unresponsive by her parents—she was breathing and had a heartbeat, but her brain was clearly not working right. They rushed her to a hospital, where she underwent a blood test to look for an infection, followed by a spinal tap to look for infected cells or bacteria in her spinal fluid. The nurses and doctors remarked on how these tiny needle punctures led to substantial bleeding. What should have been simple tests left her bruised multiple times, but they also helped lead the doctors to a critical question: "Did little Olive receive a vitamin K shot at birth?" Her parents remained quiet. But they knew that she had not. They had delivered her at a birthing center, and although the vitamin K shot was offered, Olive's parents had declined. They were devastated, and rightly so. Olive was rushed to the CT scanner, which showed a large bleed had filled half her brain. Such spontaneous bleeding would not have gone rogue had she received the vitamin K shot. Now she was in a serious predicament, compounded by the needle punctures, as her blood could not clot properly.

The choice then was for Olive to receive an emergency intravenous infusion of vitamin K and wait to see if the brain bleed would clot, though the bleed could progress during the infusion, or to undertake emergency brain surgery during the infusion, potentially putting both sides of her brain at risk for further bleeding. Her parents decided to go ahead with the risky surgery, and as the surgery was taking place, the vitamin K infusion started to kick in, and Olive's blood began to clot. Olive survived the surgery, and the outcome was good—with no long-term deficits to her central nervous system. But what a trauma she and her parents had gone through, all after the choice to not receive a shot that could have prevented this.[11]

The other nonhyped, although confusing, vitamin that could reasonably be added to the diet as a supplement is vitamin D. Humans usually get enough vitamin D through the diet as well as from sunlight, as UV rays are absorbed in the skin and a vitamin D precursor molecule is broken down to the active form, which acts as a hormone. Newborns and infants may lack vitamin D, especially those who are partially or exclusively breast-fed. Breast milk does not contain high enough levels of vitamin D, and the American Academy of Pediatrics recommends that infants not taking formula (which has adequate vitamin D) take supplements. This vitamin is critical for calcium absorption by the bones, leading to long-bone strengthening and growth. Older females may also benefit from supplemental vitamin D. Women over fifty are at increasing risk of osteoporosis, or weakening of the bones, leading to higher risk of fractures. Many physicians recommend these women take vitamin D combined with calcium to minimize loss of bone density and thus fractures. Some evidence shows that this combination also reduces the risk of colon cancer in postmenopausal females, although this data is mixed, and more likely than not, these supplements are not protective.[12] The current U.S. Preventive Services Task Force guidelines *do not* recommend vitamin D and calcium supplementation to prevent fractures in either pre- or postmenopausal women. There is not enough data to support the recommendation, and not enough is known about the long-term risks of taking these supplements.[13]

For a while, every woman over age forty was found to be deficient in vitamin D.[14] Several forms of this vitamin circulate in the blood, but the test doesn't look at all of them (yet). Several ethnic groups, primarily African-American women, have what seem to be low levels of vitamin D on blood tests, with no bone loss. If anything, this population has stronger bones and higher bone density than their Caucasian counterparts. African-Americans evolved in equatorial areas and were exposed to more sunlight per year than Caucasians—who evolved at higher latitudes—and developed a mechanism to bind less vitamin D from sun exposure, so as to prevent vitamin D toxicity. So, African-Americans bind less of the substance with sun exposure, but have normal levels in the blood of other forms of the vitamin, which cannot be analyzed on current routine blood-screening tests. Do these women need supplements? Some physicians feel that it can do no harm, so they continue to recommend vitamin D, as if there were a true deficiency. Others recommend holding off. The jury is still out on this one.[15]

While we began this chapter hounding on Linus Pauling and other false claims regarding supplements, one other vital supplement, especially for women of childbearing years, is folic acid, or folate, one of the B vitamins. Low levels of folic acid can result in a type of anemia (low blood count), though this is easily reparable. But crucially, the use of folic acid, primarily during pregnancy, has been found to dramatically reduce the incidence of midline congenital defects, including neural tube defects such as spina bifida and anencephaly (severe loss of brain tissue), heart abnormalities, and cleft lip and palate. Currently, the U.S. Public Health Service, the Centers for Disease Control, the National Academy of Medicine, the U.S. Preventive Services Task Force, and the American College of Obstetricians and Gynecologists recommend 0.4 to 0.8 mg/day of folic acid supplementation prior to and throughout pregnancy.[16] That's 400 to 800 micrograms (mcg), which is how it's usually listed on prenatal vitamins.

While some vitamins are, indeed, vital for life, and some vitamins will prevent ailments, most supplements are not necessary. But they are not harmful either, with one big, giant caveat. The majority of supplements are sold over-the-counter, but are regulated as foods (by the FDA) as opposed to medicines. The FDA regulates both nonprescription and prescription drugs, and the approval process for these has multiple levels, with the Center for Drug Evaluation and Research requiring that an exhaustive IND (Investigational New Drug) application be filed for all drugs, followed by clinical trials, before getting close to hitting the market. Supplements, on the other hand, need not go through this process at all.[17] The FDA may pull a substance off the shelves if it's found to do harm, but the supplement does not need to go through any other regulatory agency or clinical study before it gets to the shelves in the first place.[18] Labeling, including dose of a given substance, is not regulated either. This is even more concerning for herbal supplements that don't fall into any class of vitamins—such as Saint-John's-wort, saw palmetto, and ginkgo biloba. These seemingly safe "natural" supplements may have serious side effects, such as interfering with the action of prescription medications, or thinning the blood. Most surgeons are more likely to cancel a surgery on a patient who's been taking a list of supplements than on a patient who's been taking prescription medications with known mechanisms of action.

Vitamin E, for instance, can increase bleeding risk, as can vitamin C. Many herbal supplements can interfere with anesthetics and wound healing. I also hear all too often from patients, friends, and even colleagues how certain vitamins or herbal supplements changed their life—"Vitamin B_{12} gave me my energy back," "My son's focus is so much better when he takes his omega-3 supplements," "I can't get through the cold season without my daily vitamin C." Well, guess what—and this may surprise you, given my skepticism of the supplement industry—these folks are probably right. If you think that vitamin B_{12} gives you energy, it probably does. If your son seems more focused with omega-3, then great. And if you get fewer colds when you take your daily C, then keep at it.

Most vitamins, taken without megadosing, are harmless, but the placebo effect that goes along with taking them is more powerful than most can imagine. The placebo effect is not fake—it is a real physiologic reaction that the body has when something people feel they do will make that change. This has been examined in countless drug studies, and, before stricter guidelines on human subjects were created, in sham surgery studies. I'm not suggesting that you should go out and take supplements because you can trick yourself into making them help you, even if you know they won't. But I am saying that, for the most part, supplements taken in moderation (for instance, one multivitamin per day instead of seven) are less likely to harm you than help you. This is because people take supplements as part of a healthy lifestyle, and all that goes along with this, including eating well, exercising, and getting enough sleep. They're simply part of the package. The science may show that vitamin X does not prevent disease Y, but if you feel that your vitamins help, then go for it.

More likely, your vitamin intake simply produces some of the most expensive pee and poop on the market. While vitamins are critical, and we've established that they are needed in one's diet, the overwhelming majority of us get more than enough in a regular diet—be it Paleo, gluten-free, or even vegan. Even the pickiest toddlers get surprisingly sufficient amounts of vitamins in their diets. Parents commonly complain of color eaters—"He only eats white food" (plain pasta, milk, rice, and string cheese). Well, pasta contains vitamin B, including folic acid, and iron. Rice contains vitamins B_1, B_2, B_3, and B_5. Milk and cheese contain vitamin D and calcium, with a fair amount of protein. Throw in some Rice Krispies, albeit not the healthiest choice, as it's loaded with sugar, and you've got some C, E, and A for that incredibly picky eater. For those of us who have a more colorful diet, vitamins are everywhere. Certainly more naturally occurring vitamins are in fruits and vegetables as opposed to processed foods fortified with vitamins, so anyone who has fruits and vegetables as part of his or her regular diet is getting sufficient vitamins.

While some take supplements, I firmly believe that even though a prescription is not needed, these should each be discussed with one's doctor. We can't assume that, just because one doesn't need a doctor to order the medication on a piece of paper, it's not worthy of both disclosure and discussion. If a patient has had kidney or liver problems, a doctor needs to advise which vitamins should be avoided. If a patient is on certain prescribed or over-the-counter medications, the doctor needs to know so as to advise which herbal supplements should be avoided. If a patient is to undergo surgery, the surgeon needs the whole list, including supplements and herbs. Many patients don't consider vitamins or herbal supplements medications, so they often leave this critical piece out of the puzzle. I can't tell you how many times I've asked, "What medications do you take?"

And the answer is "None."

"No vitamins or supplements?"

"Oh, of course I take a multi, two D's, ginkgo, fish oil, and a powder I received from my holistic specialist. But I don't take any medications."

Sigh.

HYPE ALERT

* Vitamins are among the most overly hyped products today in health circles, but even without them vitamin deficiencies are extraordinarily rare.

* Extra vitamin C will not prevent or reduce the length of a cold.

* Some vitamins are critical during certain periods in one's life. Vitamin K, for example, is a lifesaving shot given to newborns to prevent spontaneous bleeding. Folic acid is important for pregnant women to take to prevent neural tube defects in their babies.

* Because the vitamin and supplements industry is largely un-regulated, you don't always know what you're getting.

* All vitamins and supplements taken should be discussed with your doctor as if they were bona fide medications.

8

RAISE YOUR GLASS: WATER, WATER, EVERYWHERE

How to Be Smart Without Drinking Smartwater

Do we need to drink eight glasses of water a day?

Does drinking water make your skin look younger?

Why are male surgeons prone to kidney stones?

Which is better, bottled or tap?

Caffeine and alcohol—how much is too much?

Our planet's surface is nearly three-quarters composed of water. So are we. Water is such a critical resource that its presence is literally a matter of life and death—for countries, cities, and individuals. Linda Sue Park's bestselling book *A Long Walk to Water,* in part about a young South Sudanese girl's daily four-hour trek to fetch water for her family to survive, has brought even more national and international recognition to the critical value of something we all too often take for granted. California was in a serious drought for over ten years, with water restrictions on individuals getting real. My kids learned that every time they didn't flush the toilet (when it's yellow, let it mellow), they'd save three gallons of water for our parched state.

Water is both a real and a metaphorical representation of survival. Indeed, water is "all that." But its powers are limited. We need water to survive, and we need it to thrive. But while its nourishing powers are more than skin-deep, it does not flush out toxins or give radiant skin or prevent wrinkles. While even in medicine a crude, yet valid, measure of a patient's hydration is skin turgor (lax skin signals dehydration, and firm skin shows good hydration), much of the skin's turgor has to do with its collagen, elastin, and subdermal fat, all of which, water or not, wane with age. So while water is critical for survival, it won't give you younger skin.[1]

Back in 2007, the *British Medical Journal* published a report from researchers at the Indiana University School of Medicine who created a list of common medical beliefs embraced not only by the general public but also by physicians.[2] These beliefs are either totally false or lack scientific support.[3] The number one myth was that we should be drinking at least eight glasses of water a day. No doubt you've heard this recommendation repeatedly. The researchers, Dr. Rachel C. Vreeman and Dr. Aaron E. Carroll, who went on to write books debunking medical myths, found no scientific data to back up this advice. They did find several unsubstantiated pieces of advice in the popular press, and when they tried to pinpoint the instigator of this tip, they went all the way back to a 1945 article from the National Research Council, which is now part of the National Academy of Sciences. The article noted, "A suitable allowance of water for adults is 2.5 liters daily in most instances. An ordinary standard for diverse persons is 1 milliliter for each calorie of food. Most of this quantity is contained in prepared foods."

If you ignore the last part of that comment—*most of this quantity is in foods*—then you may get the impression that you need to drink 2.5 liters (84.5 ounces, or roughly eleven eight-ounce glasses) of water a day. Not so. It's easy to obtain most of the water you need in a day without actually drinking it. But this requires that you watch your diet, which is a good thing. Water is the most prevalent ingredient in whole fruits and vegetables. Pineapples, oranges, and raspberries, for example, are 87 percent water by weight. Nudging a bit higher are peaches with 88 percent water, cantaloupe with 90 percent, and grapefruit with 91 percent. Plums and blueberries contain 85 percent water. Apples and pears are 84 percent water. Strawberries and watermelon are king, being about 92 percent water per volume. Bananas may seem dense but are 74 percent water.

Among the vegetables, those with the highest water content include cucumber and lettuce (96 percent); celery, tomatoes, and zucchini (95 percent); cauliflower, red cabbage, eggplant, spinach, and peppers (92 percent). Broccoli clocks in at 91 percent water by weight, and the classic white potato comes in at a surprising 79 percent. Meat, poultry, and dairy have water, too. Baked salmon is 62 percent water; a roasted chicken breast is 65. Even cheeses such as cheddar and blue are about 40 percent water. Grains, beans, and pasta soak up a lot of water when you cook them, which is why one cup of couscous supplies half a cup of water and a cup of red kidney beans is 77 percent water. The idea that caffeinated drinks, such as coffee and tea, are dehydrating isn't necessarily so. The caffeine does act as a diuretic, but the water in the drinks makes up for the loss.

Why has the eight-glasses-a-day myth persisted? Once again, we need only follow the money—many of those encouraging the public to overhydrate have financial ties to the bottled-beverage industry. In 2016, bottled water overtook soda as the number one drink in the United States by sales volume (and indeed Coca-Cola and PepsiCo are the leaders). While bottled water is healthier than bottled soda or juice drinks, the line is becoming fuzzier between these highly marketed beverages. While Smartwater is just water (and who wouldn't want to look like their celebrity spokesperson, Jennifer Aniston?), albeit expensive water with unnecessary ingredients and plastic containers that may end up littering the environment by ending up in our oceans, many of the bottled-water products, especially those owned by soda companies, are heading into the sugar-filled-soda category. Sparkling and flavored forms of "water" may contain nearly as much sugar as soda.

HOW MUCH *SHOULD* YOU DRINK?

Before I answer that question, let me share water's killer impact if you take in too little or too much. When I was a younger surgeon, my hos-

pital operating days were always on Fridays. Many of the Friday surgeons treated the sickest patients of the week, so as not to delay care
over the weekend. Patient transfers from outside hospitals are often on
Friday afternoons, and it's not uncommon for people to wait to get to
a hospital until the end of the workweek. For these and many other
reasons, my OR Fridays ran notoriously late. I would commonly perform eight or ten airway operations on a Friday, many of these on
critically ill patients. On some Fridays, eating and drinking were not
priorities, but drinking always took a distant second to eating. Girl's
gotta stay nourished! Snacks in between OR cases tend to be vending-
machine treats—pretzels, crackers, or chips. I'll never forget one particular winter Friday, driving home in the dark, feeling kind of funny. I
chalked it up to being tired. And hungry. I didn't even consider being
thirsty until I peed bright red blood. I was so dehydrated that I actually
had no urine to pee. Gross, I know, but it was a major wake-up call,
never to not drink for a whole day, even when not thirsty. But time
can block out even the biggest mistakes one makes.

Years later, when this scare was a distant memory, and I had advanced in the ranks to become a Tuesday surgeon, I had an especially
long clinic day. It was a Monday. I remember the meal with my husband that night, and I remember being especially tired. This time I
chalked it up to being twenty-eight weeks pregnant with my first child,
in addition to a long workday. I went to bed early, but was woken at
about midnight by a strange feeling. Contractions. Terror. It's too early
to have this baby. When we got to the hospital that night, and the ob-
stetrician on call confirmed that the baby was fine, and that I was not
in labor, my husband and I both stood up and headed for the exit. The
next day was a surgery day, after all, and we couldn't miss that.

"Not so fast," the doctor, who seemed half my age, declared. "You
still need to be monitored overnight. And besides, these contractions
began because you're extremely dehydrated. Drink more water, and this
probably won't happen again until your baby is due." Argh! I knew I
had forgotten something. I didn't drink all day and hardly had much
liquid during dinner. I spent the rest of my pregnancy carrying (and
drinking) water wherever I went. My OB had got news of this late-night

hospital visit and banned me from the OR the next day. I argued that a newborn was in the ICU with breathing trouble and she needed a tracheotomy, and the surgery could not wait. My OB said, "Find someone else to do it." So I did.

The surgery went fine, and this girl, with multiple chronic congenital medical problems, is still my patient. When I see her, now a teenager, I am still reminded of the fragility of one's life as well as the lives of others. I still feel a bit stupid that I couldn't do her surgery because I was foolish. If we don't take care of ourselves, we're of no use to others. And it's not just we female surgeons who have a touch of masochism in our dehydrated blood. One of the most well-known occupational hazards of being a male surgeon is the development of kidney stones. The combination of not drinking for long periods, standing for long periods, working under hot lights, and eating salty snacks in between surgeries leads to literal sludge in the kidneys. This builds up over time, and before age fifty, many male surgeons have had at least one kidney stone. In general, men are at higher risk for kidney stones than women, regardless of occupation. When our department consisted of just a small group (five or six) of macho men, the kidney stone rate was a solid 100 percent.

Now let's consider the other side of the coin: too much water, which is a more common problem that you probably imagine. A friend of mine is an ultramarathoner. This unique subset of self-declared lunatics push themselves beyond marathon running. These people knowingly travel to the roughest terrain at the highest altitudes. Races can be in deserts, across mountain ranges, or in jungles. Distances range from fifty to one hundred miles. These are superhuman feats, which is one of the reasons only a small, select group partake in these incredible events. The physiology of these beings has been well studied, including how they acclimate to altitude and climate variations, how they train, and how they survive. One of the key points my ultramarathoner friend is consistently focused on is water intake—not too little water, but too much. "Water kills," she definitively states. And she's not far off. With any exercise, let alone extreme exercise, fluid and electrolytes such as sodium, chloride, potassium, and bicarbonate must be in perfect balance. The losses not sensed—not from peeing and pooping, but from sweating and breathing—are not purely water, so replacements must match the losses, which include electrolytes. Throw in altitude change, weather, and degree of exertion, and the fine balance becomes ever more complicated.

Yes, water is important, but just as essential are salt, other electrolytes, and complex carbohydrates, both before, during, and after endurance exercise. If an ultra athlete in action drinks straight water, it can lead to intractable nausea, vomiting, and, surprisingly, extreme dehydration. It can be deadly. My friend will never forget the time she was on an ultrarace in the Swiss Alps. It was windy, and her life-sustaining packet of electrolyte powder to mix with just the right amount of water got blown away in the wind. It was her last one. She was left with two options: drink straight water, or no water at all. She chose the latter, which was the smarter choice. She knew that drinking plain water wouldn't kill her, but it would trigger such intense nausea and vomiting that she wouldn't be able to finish the run. Because a race volunteer had a few extra packets, it all worked out just fine and my friend finished strong.

BOTTLED IS NOT BETTER, EVEN IF IT'S INFUSED

What about bottled-water brands that claim health benefits, such as water that is ionized, alkalinized, oxygenated, or supplemented with vitamins? Or those that claim to be pure and clean—from some gently flowing stream in a meadow or down the side of a mountain? Before the advent of the bottled-water industry in the early 1980s, if you'd told your neighbor that he would soon pay more for water than gasoline, he would've laughed. Fast-forward a few decades, throw in media messages questioning the safety of tap water and a handful of bottled-water celebrity endorsements, and you arrive at today's $12 billion water industry, one of history's greatest marketing coups.

Bottled water costs two thousand times as much as tap water. The irony is that most bottled water is tap water, only less healthy, as studies have shown that harmful chemicals in the plastic can leach into the water. In the final years of the twentieth century, and well into this millennium, plastic water bottles contained BPA, or bisphenol A, with over 6 billion pounds produced and more than a hundred metric tons released into the atmosphere each year. While bottled water is no longer packaged in plastic containing BPA, much of the damage has already been done. And as I called out in an earlier chapter, other bisphenols (e.g., BPS and BPF) that are probably just as injurious are currently being substituted for BPA.[4] (Unfortunately, new materials can come on the market before they are sufficiently tested.)

BPA (and its analogues) are notorious endocrine disrupters. BPA was first made in 1891 and was used as a synthetic estrogen drug for women and animals in the first half of the twentieth century, until it was banned for medical use for its cancer-causing effects. In the late 1950s, BPA began to be used in plastics after chemists at Bayer and General Electric discovered that it could form a hard plastic when linked together in long chains (polymerized). It soon found its way into everything— from electronics, cars, and food containers to dental sealants and cash-register receipts. We don't consume these things directly, but the chemical can leach from food containers and bottles, entering their contents that we do consume.

Bisphenols affect the endocrine system because in the body they mimic estrogen. BPA has been implicated in contributing to earlier onset of puberty in girls, obesity, and several types of cancer.[5] Bottled water is also an environmental nightmare. Over 5 trillion plastic pieces, weighing a total of 250,000 tons, are floating in our seas worldwide. This has created hazardous waste due to the degradation of these plastics from heat and erosion, with devastating effects on wildlife. Marine life has also been negatively impacted via direct entanglement and ingestion.[6]

Tap water is better regulated than bottled. Bottled water is monitored by the FDA (Food and Drug Administration), while tap water is monitored by the EPA (Environmental Protection Agency). In monitoring bottled water, the FDA does not need to disclose sources of water, treatment processes, or contaminant reports, while in monitoring tap water, the EPA is required to send an annual water quality report to local residents disclosing this information.[7] In the early 1990s, when bottled water became more and more in vogue, pediatric dentists started to see an increase in cavities. Tap water sources include traces of fluoride, which, in the appropriate amounts (in tap water or fluoridated toothpaste), has been instrumental in reducing cavities in young kids. Bottled water filters out fluoride, so the kids drinking bottled water alone weren't getting the same protection as their tap-watered classmates. But as with most deficits from a new market, the bottled-water industry found yet another new angle and started selling "bottled water with fluoride," as if this were a bonus, with kid-friendly labels aimed at children. These companies claimed that their water was better with the addition of fluoride, when, in fact, they just weren't having it removed.

That said, tap water can be problematic in some communities. When high levels of lead were found in Flint, Michigan, in 2015–16, everyone took notice. High lead exposure, especially in young children, can lead to neurological deficits and anemia. But the Flint, Michigan, event was isolated and rare. For the most part, tap water supplies in the United States and other developed countries are safe, and more closely monitored than bottled-water supplies. Adding water filtration systems, either to your tap supply or via a filter attached to a pitcher, usually does not filter out fluoride.

Many of us enjoy carbonated water, either flavored or otherwise. It doesn't purport to have superpowers—it just tastes good, has no sugar, and no calories. Is it worse than tap water? It's probably about on par with any other bottled water—no better, no worse. No, despite popular claims, it does not leach out calcium from the bones, nor does it rot your teeth. The carbonation does have a by-product called carbonic acid, which may have some mild effect on tooth enamel, but this is nowhere near the level of what sugary soda can do. I remember a trick we were threatened with as kids—drop a baby tooth in a glass of Coke overnight, and it will have dissolved by the morning. I don't know anyone who tried this, as we all preferred going the way of the tooth fairy, so I have no hard evidence that teeth rot from soda.[8]

Gatorade, the most popular enhanced water of all time, is no more beneficial to the average athlete than water. It contains fourteen grams of sugar per eight ounces (most bottles of Gatorade are twenty ounces, so that's a whopping thirty-five grams of sugar per bottle), and since it's now being used more and more often by young kids who aren't even athletes, it's become yet another culprit in the rise in childhood obesity and early-onset diabetes. Blk. water (water combined with fulvic acid, which turns it black) came to be after it "miraculously" cured the inventor's mother's cancer. Not only is fulvic acid worthless in improving health, it apparently doesn't even taste good—one of the product's more colorful Amazon reviewers compared it to sucking on rotting leaves from the bottom of the Hackensack River.

Vitamin-infused water is another illusion. I have a physician friend who teaches health education in schools. A fan favorite is her vitamin-water class ("Okay, guys, today we are going to *make* vitamin water!"). Twelve-year-olds cheer. She then pours sixteen teaspoons of sugar into a two-liter bottle and adds water and a few drops of food coloring. "Okay, drink up!" Surprisingly, some actually do. While vitamin water does, indeed, contain vitamins, the majority of the contents are sugar and dyes. Even some of the vitamins it touts, including vitamins A, D, E, and K, are fat soluble and are not necessarily absorbed at all from a drink, but are absorbed into your bloodstream through the fat you eat. And the water soluble ones, vitamins B and C, are just excreted (peed out) when taken to excess. Most of us get from food substantially more than the U.S. recommended daily allowance of vitamins, making vitamin-supplemented water all the less necessary. Yes, vitamin water and other supplemented-water products are healthier than sodas, containing less sugar and fewer calories, but not by much.

Part of the problem with supplemented water is that people associate ingestion with absorption. This is far from how things work. If you drink amino acids, they don't miraculously bond into usable proteins, and even if they do, they don't necessarily get absorbed directly to your muscles (no, there is no such thing as muscle food or brain food). Because we surgeons literally see the internal organs sweat (what's called insensible losses), we appreciate how things really get replenished. Undergoing an open (as opposed to laparoscopic) abdominal operation can be just as stressful on the body's fluid/electrolyte balance as running an ultramarathon. So many factors must be taken into account: NPO (*nil per os,* or "nothing by mouth," or no eating/drinking) status before, during, and after the surgery; duration of the surgery; blood loss; recovery time until eating/drinking; health status going into the surgery, including heart, lung, kidney, and liver function. Surgeons and anesthesiologists, as well as medical specialists on the pre- and postoperative sides, are accustomed to factoring all of these issues into fluid and electrolyte replacement, both based on what's known in advance, as well as in real time as things happen.

This incredibly delicate balance on a body that's stressed in ways one cannot imagine can be done, almost to the drop of water, and to the molecule of salt. But in this arena, we are talking about *intravascular* hydration, also known as intravenous hydration. One cannot "drink" the type of fluid that is lost in a big surgery and expect to have it go where it's needed. This is true for water or for any beverage that has supposed supplements, vitamins, or superpowers. Anything that goes through the gastrointestinal tract undergoes *some* absorption into the vascular system, but most just gets excreted via the kidneys as urine or via the colon as stool.

CAFFEINATED

You don't want to interact with me before I've had my morning coffee. I may not be such a great person afterward, but I'm pretty uninspired without it. Many people will say this about themselves. Caffeine is the most socially acceptable addictive drug on the market. Surgical training only made this more of a reality for me, and for so many others who function on too little sleep. It's not that we're superpowered after our cup of joe, but we're pretty inert without it. And unlike alcohol or other drug addictions, caffeine addiction remains without social negativity. Nonetheless, it is a drug, and we're addicted. That said, it's not such a bad drug, and despite its addictive nature, our daily caffeine buzz can actually be to our advantage. Some studies have found habitual caffeine consumption to reduce risks of Alzheimer's disease, Parkinson's, and certain cancers. But it may be associated with increasing risk of bone loss and high blood pressure. Excessive caffeine intake during pregnancy—more than 300 mg per day—has even been associated with fetal loss.[9]

Despite ongoing studies, and multiple large population analyses, caffeinated-beverage consumption continues to be a mixed bag: it's good for you; it's bad for you. It stunts your growth; it helps you concentrate. Don't drink caffeine while pregnant; it's okay to drink caffeine while pregnant. While coffee continues to be the most commonly consumed caffeinated beverage, caffeine is present in sugary carbonated soft drinks, teas, energy drinks, sugary and fat-filled coffee and tea drinks (pass the mochaccino, please), and chocolate products. Adults tend to be the highest consumers of coffees and teas, and kids and teens are the ever-increasing drinkers of all of the other caffeinated treats. Even those that are not billed as such will likely contain some caffeine, sometimes even more than the products with the "energy" label. Have a headache? Take some Excedrin and you'll get a bonus 65 mg of caffeine—nearly as much as in a good strong cup of coffee.

Eighty-five percent of the U.S. population, including children as young as age two, consume at least one caffeinated beverage per day. That's right—two-year-olds drink caffeine-containing soda and chocolate drinks, and some are even given coffee. One cup of regular coffee contains approximately 95 mg of caffeine, and the daily consumption of up to 400 mg has been found to have no negative impact on health—it's not necessarily good, but it's not bad, either. For kids ages six to twelve, that drops down to 45–85 mg/day, and for pregnant women, 200 mg or less. So, yes, one (large) cup of regular coffee per day is fine for pregnant women. This is all according to the U.S. FDA recommendations. But they are not touting caffeine's benefit; they are simply saying that it is not harmful.

One can of Red Bull contains 80 mg of caffeine, about as much as one cup of coffee.[10] Caffeinated soda contains about 5 mg/ounce, but since servings are usually twelve ounces instead of eight ounces, each can packs about 60 mg of caffeine, almost as much as a cup of coffee. But for those of us who buy coffee in commercial chains such as Starbucks or Peet's, it is rare for one to see those eight-ounce cups, unless it's for a large espresso shot or for serving a small child. Most baristas offer small (Tall), medium (Grande), or large (Venti), equal to twelve ounces, sixteen ounces, or twenty ounces. Gone are the eight-ounce cups once seen at the diner counters of yesteryear. And if the 330 mg of caffeine in a Venti isn't enough for you (it certainly isn't enough for me), the options get better—a Red Eye is a coffee serving of any size with an added shot of espresso; a Black Eye adds two. So a Black Eye Venti at Starbucks packs about 540 mg of caffeine. In some situations, up to 1,000 mg of caffeine per day is considered safe (according to the FDA), for instance, for military personnel who need to be awake for longer periods. But this statement can be fraught with a slippery slope of danger. Do new parents need to be more awake to get through sleep deprivation from a new baby? Do physicians working long shifts need extra jolts? What about the pilot flying your plane? Or your teenager cramming for a test? That 1,000 mg is more than ten cups of coffee. This just can't be okay for our health.

Overdosing to the point of toxicity when it comes to caffeine is not that easy to do. Sure, most will feel pretty crummy, perhaps with some jitteriness, stomach upset, or rebound headache after more than four or five cups of coffee over a relatively short period. Although it takes a lot of caffeine to call it a lethal overdose (you would need to drink around 14,000 mg of caffeine, or 140 cups of coffee), that astronomical number doesn't mean it's fine to gulp three cups of coffee per hour. In some people even much lower doses of caffeine can lead to heart arrhythmias, dangerously rapid heart rate, high blood pressure, or sleeplessness. Some of these can cause a heart attack, stroke, or death, especially in people with previously undetected abnormalities of their cardiovascular system.

What's more concerning are the beverages with hidden caffeine, or other similar stimulants. Caffeine is found in an expansive array of foods and beverages these days, and you can consume exceedingly high

doses all at once with added jolts from chemicals such as tyrosine, tau-
rine, and phenylalanine. Not only will these make you hyper, but at
the right dose they can put you in the hospital with a hyped-up heart.
Many energy drinks do not disclose the amount of caffeine in them,
mainly because the FDA does not mandate it. In 2012, *Consumer
Reports* reviewed the twenty-seven most popular energy drinks.[11] Not
surprisingly, their range of caffeine was large—from as little as 6 mg
(about 5 percent that of an eight-ounce cup of coffee) per serving, and
some contained more than one serving. Even less surprisingly, the
Decaf 5-hour Energy shot contained the least amount of caffeine. The
"Why Bother" energy shot, in other words. Disconcertingly, eleven of
the twenty-seven products do not include the amount of caffeine per
serving on their label, even though they are clearly being marketed as
energy-boosting, caffeinated drinks. And some that do provide the caffeine content are giving a gross underestimation.

Caffeinated beverages are no longer limited to those who need a
morning boost, an afternoon pick-me-up, or a late-night nondrowsy
aid. Athletes have turned to caffeine to enhance performance. A stim-
ulant, it acts as a central nervous system energy enhancer. It's the world's
most popular (and legal) psychoactive drug. It still remains legal in the
sports arena, according to the World Anti-Doping Agency (WADA).
More and more athletes have been using caffeine to improve their race
speed, and even overall performance in nonracing, stop-and-start sports
such as tennis or volleyball. Some athletic organizations, screening over
twenty thousand urine specimens for illegal performance-enhancement
drugs, have found caffeine in as many as 75 percent of the samples.
While caffeine won't make a nonathlete into a champion pole-vaulter, it
can give a kick start if used in moderation. And as with any drug, which
caffeine certainly is, there's a fine line between therapeutic and making you feel awful. The sweet spot for caffeine's boost for athletic
performance is 3–6 mg/kg body weight. So for a 70 kg (150 lb) athlete,
that would be about 220 mg of caffeine, or a medium Starbucks coffee.
Anything above 6 mg/kg can lead to jitteriness, abnormal heart rate or
rhythm, or high blood pressure. And again, it won't dehydrate, contrary
to the persistent myth of caffeine's effect.[12]

As with any performance enhancer, the slope becomes slippery with the only-human notion of more is better. WADA does not classify caffeine as an illegal enhancement drug, but I wouldn't be surprised if this changes in the future. It was a banned substance until 2004, when the criteria were reevaluated. To be banned a substance must meet two of the three criteria: enhances performance (check), poses a health risk (WADA unchecked this box, but I'd question that), and it violates the spirit of the sport (no way—coffee is a social beverage!).

The seeming benefit of caffeinated products to keep you more awake and alert during the day, particularly when consumed too late for your body to properly metabolize, can lead to sleep deprivation and exhaustion the next day. Adolescents and young adults are spending more and more late nights on a screen, and they use caffeinated products (not just coffee) to stay awake until the wee hours of the morning.[13] The not-so-surprising result is poor performance at work and school the following day, including dozing in class or, worse, behind the wheel. The negative physiological effects from caffeine overuse, such as high blood pressure, racing heart, or cardiac arrhythmias, can happen in teens just as often as in adults.

WINE-OH!

We often hear that wine, especially red, contains "healthy" ingredients so long as it's consumed in moderate amounts. What does that mean? The real benefits to drinking wine have less to do with its chemical makeup and more to do with its relaxing properties, making it a stress reducer. So long as you limit your consumption (specifics on this shortly). Just recently, researchers examining the causes of liver cancer calculated that it takes three drinks a day to risk giving yourself the disease. This wasn't a small study. The researchers analyzed thirty-four studies involving 8.2 million people and more than 24,500 cases of liver cancer.[14] They also found more evidence to salute coffee, which seems to protect people from liver cancer. (But that doesn't mean you can drink three glasses of each beverage and cancel out your risk. The "bigger" culprit to liver cancer that few people think about is excess weight. Being overweight or obese can cause liver cancer, among many other types of cancer.)

Wine must have some benefits to merit its being touted by so many studies as well as health professionals. What's its magic potion that has led to its being considered healthy, and even protective from many illnesses? One of the longest-running studies was the Nurses' Health Study (NHS), conducted over thirty-two years (1980–2012) in an effort to evaluate the impact of health practices over many years for many people.[15] One of the factors evaluated was the benefit (or risk) of drinking alcohol, and how that would impact long-term health risks such as cancer, cardiovascular disease, fractures, and mental illness. The study's initial recruitment began in 1976, to include one hundred thousand nurses. Given the date of the original NHS (there have been several more over the following decades), nurses meant women. Its counterpart, the Physicians' Health Study (PHS), involved men at around the same time, but specifically evaluated the impact of aspirin and beta-carotene.

What was so remarkable about the NHS was that it followed people via questionnaire over decades, including such factors as cigarette smoking, alcohol use, and exercise habits, in efforts to link these behaviors to risk of breast cancer, cardiovascular disease, other cancers, fractures, and mental illness. Close to three hundred thousand people have been enrolled in this study since its start, and it now includes both men and women.[16] No health recommendation is going to be cut-and-dried, and this goes for alcohol consumption. The NHS study found that women drinking one glass of wine (or men drinking up to two glasses of wine) more than three times a week is more beneficial than women drinking more than one glass (or men drinking more than two glasses) fewer than three times a week. In other words, drinking in moderation was found to be better if done so more often but in smaller servings. While moderate alcohol intake increases the risk of breast and colon cancer, as well as hip fractures, it lowers the risk of cardiovascular disease and neurocognitive deficits.

The component resveratrol has been all the rage as an added health benefit from red wine in particular. This substance has been shown to have some magic health power, so much so that it now comes in pill form.[17] First of all, what is it? Resveratrol is a polyphenol naturally contained in purple grapes, blueberries, cranberries, mulberries, lingonberries, peanuts, and pistachios. In 2012, an article in the journal *Cell* demonstrated, on a biochemical level, how this substance may increase mitochondrial function, physical stamina, and glucose tolerance in mice.[18] It was also shown to ameliorate diet-induced obesity. In other words, this chemical could be the answer to diabetes, obesity, and age-related illnesses. The media went wild over this, even though the effects were in mice, and even though these results did not initially apply to humans at all. Nonetheless, we now had a flowing red fountain of youth. But no study in humans has yet shown direct health benefits from this substance. As with most studies that measure health outcomes based on behaviors such as wine consumption, obtaining any useful information requires years of prospective data with large populations. It can't be done in a lab and certainly can't be done on animals. Animal studies provide useful information on which to base human studies, but equating the two is not accurate.

Beyond the resveratrol quandary, alcohol consumption in general has been found to be good for your health. With an enormous caveat: in moderation, meaning no more than one glass before or during the evening meal. Resveratrol or not, moderate habitual alcohol consumption has been associated with lower mortality, lower cardiovascular risks, lowered risks of diabetes, heart failure, and stroke. However, if you're not a drinker, don't feel you need to start. There is no evidence that drinking alcohol is cardioprotective as a rule. The studies have demonstrated *associated* lower risks of certain chronic illnesses, but not a direct cause and effect of improved function of your ticker if you drink. Perhaps even more important to remember is that alcohol overuse is the third leading cause of death in the United States and is the leading cause of premature death in fifteen- to fifty-nine-year-olds due to both medically related issues such as liver disease as well as to alcohol-related accidental deaths.[19] So when you are considering the benefits of alcohol consumption, drink lightly. No studies have shown that drinking alcohol of any kind is a direct cause of better health, but countless studies have shown absolute direct causes of negative effects. Excessive consumption of most alcoholic beverages is a leading cause of obesity, liver disease, accidental deaths, and even domestic violence.

The beverage industry continues to evolve. No longer is it a question of how much water one should drink, but what kind of water it should be—bottled, tap, flavored, enhanced, or bubbly. The answer to how much can be simplified from the once-popularized "eight glasses, eight ounces each." Check your pee! Light yellow? You're good. It's a matter of hydration, which may not come in the form of water at all—fruits and vegetables count, and so does my trusted friend coffee. We tend to hang on to notions of new and improved. Quite frankly, it's hard to improve clean tap water. However, it just seems too simple, too pedestrian, to accept this. Perhaps a television celebrity can market the idea of tap water (pennies per gallon!) better than any doctor could.

HYPE ALERT

* You can get most of your water through food so long as you have a good diet.

* Sparkling and flavored forms of "water" may contain nearly as much sugar as soda.

* Bottled water is not better than tap, even if it's infused with seemingly healthy ingredients including vitamins.

* Sports drinks are not necessarily better for athletes than water.

* Traditional coffee has more potential benefits than health risks. Caffeinated beverages with added sugars and other ingredients, however, are a different story.

* Moderate drinking of alcoholic beverages may confer health benefits, but don't feel that you need to start drinking if you're not a drinker.

9

PUTTING THE *C* BACK IN CAM: COMPLEMENTARY ALTERNATIVE MEDICINE

How to Stay Natural While Taking Your Medicine

> DO ALTERNATIVE TREATMENTS FOR CANCER EVER WORK?
>
> DOES HOMEOPATHY HAVE ANY MERIT?
>
> WHY IS THE PLACEBO EFFECT SO POWERFUL?
>
> WHAT IS THE "REVERSE" PLACEBO EFFECT?

UCLA Health Center may lie just a few miles from Hollywood, but some of the things I see people do that land them in the hospital would give any screenwriter a run for his money. Within the last year alone, I've seen a man lose his voice box because he avoided simple surgery for a treatable throat cancer; I've watched toddlers nearly die in the operating room after they've choked on nuts given to them by their parents to "boost their omega fatty acids" naturally; and I've had a highly educated woman tell me that she used a bleaching solution to clean her ear canals—unfortunately this led to burns across the entire ear and face. When it comes to the shoddy medical advice some people follow, reality is definitely stranger than fiction.

But of all the wild stories I hear in the hospital that have a deep impact on me, the ones about alternative, desperate cures for cancer top the list. Let's go there first, with two cases in point that will always stand out in my mind.

A man we'll call Tom was diagnosed with a very treatable tongue tumor. He decided to go for hyperthermic therapy instead of conventional surgery. In hyperthermic, or thermal, therapy, body tissue is exposed to high temperatures (up to 113° F, which is actually much lower than in a typical dry sauna). Research has shown that high temperatures can damage and kill cancer cells, usually with minimal injury to normal tissues. But it's almost always used in combination with surgery, radiation, and/or chemotherapy. So for this patient, what could have been a well-tolerated surgery with an excellent short- and long-term outcome turned into a disastrous end-of-life story. In Tom's case, the tongue tumor continued to grow, and he developed metastases (secondary tumors) in the lymph nodes in his neck. The bulk of the tumor rendered him unable to swallow or talk because it was blocking his mouth and throat. The tumor also invaded the nerve that is responsible for tongue movement, so his tongue, which is all muscle, became paralyzed. The tumor invaded the nerves responsible for tongue sensation, so he would inadvertently bite his tongue, it would bleed, and he wouldn't feel it. Because he had no sensation and no ability to swallow, he would routinely choke on his own saliva and eventually developed chronic aspiration pneumonia and needed a feeding tube surgically placed into his stomach (gastrostomy tube).

The combination of enlarging lymph nodes in his neck and the bulky, paralyzed, numb tongue interfered with his breathing and ability to maintain his secretions, and he needed a tracheotomy tube placed in his neck. The tumors continued to grow; they could not be cut out and invaded important structures. Lymph nodes in the neck can invade major blood vessels, specifically the carotid arteries and jugular veins. Tom didn't bleed or choke to death, but his "cancer burden" became too much. He lost weight despite the feeding tube, became frail, and his immune system weakened from chronic malnutrition and the cancers in multiple systems. He eventually died of a relatively minor respiratory illness—something we'd commonly refer to as a cold.

The story of my operating-room nurse paints a similar picture. Jane was an extremely experienced and skilled nurse in her late forties, a single mom with two teenagers at home. When she was diagnosed with breast cancer, she took medical leave and returned to work about six months later. She looked amazing—vibrant and fit, slimmed down, with brighter skin. I told her she seemed to have weathered the treatment so well. Jane said, "I did alternative treatment: raw-food diet, teas, meditation, and exercise. I feel great." I was impressed, but I was concerned because, from my perspective, she hadn't treated her cancer. Yet her appearance told another story. How could this be? Well, cancer doesn't usually kill overnight. Most everyone has seen someone with terminal cancer who doesn't look that ill, and then that person dies soon after. That is what's so insidious and sneaky about the disease. It can come on slowly, then reach a tipping point where it starts to show its ugly face quickly. That's what happened to Jane. Her body did respond well to healthy living, exercise, and relaxation. Had this been a part of her overall treatment regimen, she might have endured the conventional treatment even better than most. But she had no medical treatment and soon paid the price. She lasted about two weeks at work and, a few months later, died a horrible death with widely metastatic breast cancer that had gone everywhere, including to her bones and brain. Would she have lived longer had she gone the traditional route or in combination with the alternative? Yes, it's likely. Complementary medicine has a great role in enhancing standard medical practices, but it does not work in a vacuum.

Many excellent studies in cancer medicine have documented the short- and long-term benefits of complementary medicine as an adjuvant to traditional cancer therapy (e.g., chemotherapy, radiation, surgery).[1] Patients who exercise, meditate, do tai chi, eat nutritious foods, and have good emotional support networks fare far better than those who don't. But the key word to their success is *adjuvant,* or complementary. Complementary medicine in cancer treatment and chronic-pain management has revolutionized the treatment tolerance and clinical outcomes in thousands of patients. This is not a field to bash or mock, but it needs to be incorporated as part of an inclusive medical regimen. It cannot stand on its own.

According to the Mayo Clinic, nearly 40 percent of adults have used complementary alternative medicine (CAM).[2] Unfortunately, too many people forget about the *C:* "complementary." And when people take the *complementary* out of CAM, they can end up using a wide variety of therapies with little or no scientific proof of efficacy or safety. Sometimes no harm is done, but with serious illnesses such as cancer, the results can be disastrous.

Some CAM treatment centers in locales such as Arizona and Baja California market themselves like exclusive holiday resorts, magical settings for practicing "medicine." You have to wonder if they would book many customers if they were located in Flint, Michigan, or Camden, New Jersey. When they charge tens of thousands of dollars, they force many people to waste all of their savings on quack medicine.

One lies in Gilbert, Arizona.[3] The three-person medical staff (only one of whom has a medical degree) treats no fewer than twenty major illnesses, including an "all natural cancer treatment." The website reads like a greatest hits of contemporary pseudoscience, warning against everything from table salt, nonorganic candy, and vaccines, to fluorescent lighting, cell phones, and beauty products. Advertised amenities include "the red carpet treatment" for out-of-town patients, including "a 55-inch big screen" and "brand new extra comfortable patient treatment chairs." The entire website is peppered with inspirational Christian language (ironically, the nearest highway is called Superstition Freeway).

Not too far away in Mesa is another "oasis of healing."[4] Treatments here are for cancer only and include juicing, high-dose vitamin C, colon hydrotherapy, coffee enemas, massage therapy, and infrared sauna, none of which has been proven to have any effect on cancer whatsoever. Yet they advise that "just as certain as a finger will heal, if cut with a knife, so can our bodies heal from cancer." This is where my OR nurse tried to cure her cancer. She was sold on the idea that they could target and eliminate her cancer without harming her, as well as "stimulate, balance, and enhance" her immune system so her body would "stop making cancer."

Stanislaw Burzynski's clinic in Texas is one of the most controversial.[5] An investigative piece in *Forbes* notes, "The centerpiece of Burzynski's treatment is a drug he refers to as 'antineoplastons.' They have never been approved by the FDA for the treatment of cancer, and the only way Burzynski has been able to sell and administer them is under the auspices of 'clinical trials.' . . . Despite decades of giving patients antineoplastons, Burzynski has failed to publish any human clinical trials data. . . . It's also quite outside the norm for patients to be charged tens of thousands of dollars to participate in clinical trials. This can, as you might imagine, create a bit of a conflict of interest."[6]

Dr. Frank Arguello Cancer Clinic in Baja California (Mexico) markets his treatments heavily in the United States and Canada.[7] He's made up his own science to "heal" cancer through what he calls "atavistic chemotherapy." What is atavistic chemotherapy? It's treatment derived from the hypothesis that cancer represents atavism—a recurrence of a trait or character common in our ancestors millions of years ago but which has since disappeared due to genetic recombination (hind legs on whales is one such example that's been observed in animals). But the scientific evidence doesn't support this hypothesis, and therefore, treatments designed to cure cancer based on this flawed premise are destined for failure. The premise is that if cancer cells function similarly to our ancestral single-celled organisms, then drugs such as antimicrobials (also known as antibiotics) should attack and kill these cells. His website provides dramatic pre-and post-treatment photographs and mentions that he is now eligible to obtain (although does not currently possess) a U.S. medical license.[8]

Perhaps since time immemorial we as humans have been vulnerable prey for alternative treatments and approaches that buck tradition. We want to *be* the anomaly—the outlier—the one who beats the odds and statistics, especially when it comes to an illness as scary and grave as cancer. "One of the appeals of alternative medicine is that it offers not just an alternative philosophy or an alternative treatment," writes the essayist and author Eula Biss in her book *On Immunity*, "but also an alternative language. If we feel polluted, we are offered a 'cleanse.' If we feel inadequate, lacking, we are offered a 'supplement.' If we fear toxins, we are offered 'detoxification.' If we fear that we are rusting with age, physically oxidizing, we are reassured with 'antioxidants.' These are metaphors that address our base anxieties. And what the language of alternative medicine understands is that when we feel bad we want something unambiguously good."[9]

Biss brings up some excellent points. They remind me of a story I learned long ago about a German physician who, in the second half of the eighteenth century, established his own alternative methods by coining a whole new term—one that we still use today in our vernacular, though not in medicine. He studied in Vienna, where he set up a clinic following an interest in what was called animal magnetism—the belief that animals, humans included, exert an invisible natural force that could be leveraged for healing purposes. He developed this concept into a system of treatments through hypnotism, called *mesmerism,* after his name: Franz Anton Mesmer. Dr. Mesmer claimed he could cure nervous system problems using a form of magnetism. According to Mesmer, the body's health was maintained by a proper balance of a "subtle fluid," which was the same fluid responsible for heat, light, and gravity. He said it floated throughout the universe. To create animal magnetism Mesmer focused on the magnetic poles of the body, which presumably helped direct this fluid. Under his theory, the poles had to be aligned properly to work and to maintain a correct, smooth, and harmonious flow of fluid. If the fluid balance was off, a person could develop "nervous afflictions," and he or she would need to be "mesmerized" to get the poles realigned and the fluid rebalanced.

While this may sound utterly ridiculous in today's world, Mesmer's approach took off. He generated publicity, as well as notoriety. People, educated and not, were curious. The medical and scientific community feared Mesmer; the government worried about the secrecy of and subversion from his growing group. In 1777 he was expelled from Vienna, following the scandal after his failed attempts to cure a patient's blindness were revealed. So he went to Paris and established himself all over again.

By the 1780s he had accumulated new followers and set up shops with them in Paris. These believers "mesmerized" people by claiming to locate their poles and control their fluid. They used elaborate apparatuses and tools, including iron bars that allegedly carried the invisible fluid. They referred to "mesmeric tubes" and "mesmerized water." Mesmer's popularity became part mystery, part fashion. Experiencing his mesmerism, which took place in secluded areas, became a trendy thing to do.

Dr. Mesmer didn't last long in Paris, either. Investigations commenced. In 1784 a French royal commission appointed by Louis XVI tried to find scientific evidence to support Mesmer's magnetic-fluid theory. The commissioners included Antoine-Laurent Lavoisier (the father of modern chemistry) and Benjamin Franklin. In 1785, Mesmer left Paris for London, then went on to Austria, Italy, Switzerland, and eventually to his native Germany, where he returned to a village near his birthplace and died in 1815. No matter where he went, he tried to win the universal acclaim he believed he deserved for his therapies.

In today's parlance, we'd say that Mesmer was treating psychosomatic illness—diseases or symptoms that are caused or worsened by mental factors, such as stress and anxiety. The word *psychosomatic* means mind (psyche) and body (soma). A psychosomatic ailment involves both the mind and the body. Sometimes there are only symptoms and no underlying physical disease exists. We all know that our mental state can affect how bad a real physical disease is. Mesmer made a lot of money from people's gullibility. But his story reflects others in modern times. Every day people believe in products, procedures, and health claims that are brilliantly marketed but total shams. Messages related to health, many of which can be contradictory, circulate hourly. Even the skeptics can be easily mesmerized. Separating truth from fiction, finding the line between what's healthful and what's harmful, is not so simple—especially when the (mis)information comes from someone we think is an authority or just a little smarter than us.

I could almost give Dr. Arguello of Baja points for inventiveness, as I would Dr. Mesmer, but Arguello's transgressions are more offensive: he's conning susceptible patients with real, life-threatening cancer diagnoses. We're not talking about psychosomatic illness for which there is no traditional therapy.

This is what makes these centers so infuriating to me. They prey upon people when they are the most vulnerable: facing a cancer diagnosis. People get roped into thinking that these alternative treatments are viable options as opposed to radiation and surgery. None of these treatments' outcomes can be replicated by others, which is a huge red flag. And their testimonials are just that—anecdotes. They do not reflect real data.

I never balk at a patient who relates a success story using one of these methods. Many ailments are, in part, emotionally triggered, and techniques with relaxation as a central feature may be part or all of the answer. Whose proclivity to migraine headaches wouldn't subside with better sleep, meditation, and stress relief? Medical marijuana has antiemetic and vasoconstrictive effects. These are physiological, and not purely psychological. I don't write prescriptions for pot, but I don't hold it against other doctors who do, or the people who use it to treat their conditions. Even a place as mainstream as UCLA has a Center for East-West Medicine.

Chinese medicine includes some of the more advanced pharmacological agents, with genuine physiological effects. Garlic, flaxseed, and natural soy foods can indeed help lower cholesterol in some people. So can omega-3 fatty acids, fiber, and plant compounds similar to cholesterol (plant stanols and sterols). These are natural sources of cholesterol-lowering compounds, but one can hardly call them drugs or even alternative therapies. And they are not considered to be mainstream solutions in treating high cholesterol. Even though some good, albeit small, studies have supported their benefits, for the most part they are still considered unconventional for prevention and management. It's best to think of them as additional options for addressing high cholesterol that do not require a prescription and can be easily obtained in the diet. Nothing should be substituted for proven remedies such as statins, which are much more effective and powerful.

The 2015 Nobel Prize winners for medicine were mocked for using "natural medicine." William C. Campbell of New Jersey's Drew University and Satoshi Ōmura of Japan's Kitasato University shared half of the prize for their discoveries about a novel drug therapy against infections caused by roundworm parasites. And the Academy of Chinese Medicine's Tu Youyou was awarded the other half of the prize for her discoveries relating to a novel drug therapy against malaria. These infections impact about half the world's population. Both drugs' main ingredients stem not from synthetic productions generated in a lab. To the contrary, they are examples of natural products chemistry. Their efficacious compounds are isolated from organisms that naturally produce them or similar molecules.

Now, if you were to ask a group of people on the street to define *natural products chemistry,* chances are you'll find at least one or two people hawking homeopathy. But I dare you to find someone who can define exactly what homeopathy means. Let's go there next. It's mesmerism at its best in the twenty-first century.

MESMERIZED BY HOMEOPATHY[10]

Mesmerism is alive and well today in the form of not just magnetic field therapy but homeopathy. If you had to take a guess as to homeopathy's definition, you'd probably say something along the lines of "natural" medicines that don't require a prescription. Alternative medicine often gets melded with homeopathic, which has an aura of safety in the public mind. But among the different treatments and remedies that are considered "alternative" medicine, homeopathy is the most implausible of all. In this elaborate placebo system, the "remedies" lack any actual medicine. Homeopathy was founded by Samuel Hahnemann in Germany around the turn of the nineteenth century. It was based on the idea of "like cures like" and the "law of infinitesimals." Put another way, "extraordinarily diluted products that in their original form might have caused symptoms resembling those of the illness in question are administered to patients in a highly individualized fashion."[11]

Homeopathic remedies have not lost their popularity despite modern medicine (and the scientific case against them). According to the 2012 National Health Interview Survey, about 5 million adults and 1 million children in the United States had used a homeopathic treatment in the previous year.[12] Although homeopathic drugs are so diluted that no trace of the original active ingredients remain and are generally considered to be safe, many doctors (me included) worry that even chemically inactive homeopathic preparations can redirect patients away from effective conventional remedies. What's more, dangerous consequences have been documented. Zicam Cold Remedy, for instance, is a homeopathic treatment that was found to contain high doses of zinc gluconate (also called zincum gluconicum).[13] It's the zinc salt of gluconic acid. In 2009 it was removed from the market because its intranasal use was linked to anosmia (the loss of the sense of smell).

Homeopathic medications do not undergo the same scrutiny by the FDA as medications that are allopathic—traditional drugs that work by having *opposite* effects to the symptoms. These "natural" medications can cause unpleasant side effects, interact with medications one is taking, trigger allergic reactions, and contain substances that may be either useless or, worse, harmful (such as arsenic, mercury, and lead). It's fine to drink chamomile tea to relieve an upset stomach, but you should think twice about downing other popular herbs in the form of concentrated supplements, such as feverfew, saw palmetto, ginkgo biloba, echinacea, and ginseng. Technically, these supplements are not homeopathic, but they are nonetheless revered for their natural attributes. We went into much greater detail about supplements (and vitamins) in Chapter 7. Many of these herbs can lead to bleeding, delayed healing, and bruising, especially for those undergoing medical surgery. A doctor such as me is more likely to cancel an operation if my patient has been taking herbal supplements than if she smoked a cigarette and snuck in a shot of whiskey in the waiting room. Let's get back to homeopathy in particular, which has just recently come under fire, and for good reason as you'll read about shortly.

The first time homeopathy garnered serious criticism from traditional physicians was in the 1830s and 1840s. Of all the critics, the most vocal was probably Oliver Wendell Holmes Sr., a nineteenth-century Boston physician and poet, who called homeopathy a "mingled mass of perverse ingenuity, of tinsel erudition, or imbecile credulity, and of artful misrepresentation." He also called out the potential therapeutic effect on patients: "the strong impression made upon their minds by this novel and marvelous method of treatment."[14] Clearly, he was pointing out the psychosomatic component to homeopathy's placebo effects. But Holmes didn't let mainstream medicine off the hook, either, which still espoused bleeding and emetics. Homeopathy would not go away, and even some traditional doctors believed in it.

The relationship between conventional medicine and homeopathy grew stronger during the nineteenth century, as homeopaths founded medical schools and hospitals. They called their conventional counterparts *allopaths,* in reference to a traditional doctor's focus on using drugs that have opposing effects to the symptoms. And just as allopaths could use homeopathic remedies in their practice, homeopaths could perform certain procedures typically performed by allopathic doctors—surgery included!

Not until the early twentieth century did homeopathy become suspect from an academic standpoint. This was when medicine was becoming increasingly grounded in laboratory science, which triggered a shift in homeopathy's base of support. No longer were orthodox doctors championing homeopathy in their practice. "Lay" practitioners took over promotion of the field, which only grew more popular among the lay community during the counterculture of the 1960s and onward. At that time people were simultaneously feeling animosity toward established authorities and disillusioned by conventional medicine.

Alternative healers owned the field by the 1990s, bewildering the medical establishment, which could only look upon homeopathy as a fraud that could potentially harm innocent people. Evolutionary biologist, ethologist, and notable critic of alternative medicine Richard Dawkins has said, "There is no alternative medicine. There is only medicine that works and medicine that doesn't work." He defines alternative medicine as a "set of practices which cannot be tested, refuse to be tested, or consistently fail tests. If a healing technique is demonstrated to have curative properties in properly controlled double-blind trials, it ceases to be alternative. It simply . . . becomes medicine."[15]

The lack of oversight from the FDA goes back decades. Nearly thirty years ago, the FDA issued a Compliance Policy Guide to address the homeopathic drug market. This guide aimed to create standards for manufacturing practices and labeling regarding ingredients and directions for use in homeopathic products. Only homeopaths could prescribe homeopathic drugs used for "serious" conditions. However, homeopathic drugs for conditions that would resolve without treatment could be sold over-the-counter. This maneuver not only excused the FDA from evaluating the efficacy of prescribed homeopathic drugs, but also opened the door for over-the-counter remedies to be marketed as "therapeutic."

This last point is key. Unlike dietary supplements, which were explicitly excluded from rigorous FDA regulation in 1994, homeopathic medicines that have been labeled therapeutic can be regulated to protect consumers from confusion in their marketing and their potential effects. FDA regulation of homeopathic drugs made several headlines in 2015. In March of that year, the FDA investigated what the public and physicians thought about homeopathic drugs. The agency also wanted to know whether its limited supervision of homeopathic products was "appropriate to protect and promote public health." The FDA then held a public hearing over two days featuring drug-safety experts and homeopathic-care providers and their industry representatives. By September, the Federal Trade Commission (FTC) was holding its own public hearing on the advertising of homeopathic products. The FTC wanted to know whether the industry was violating section 5 of the FTC Act, which prohibits deceptive acts or practices affecting commerce. The following year, the FTC issued an "Enforcement Policy Statement Regarding Marketing Claims for Over-the-Counter (OTC) Homeopathic Drugs," whereby the FTC will now hold efficacy and safety claims of homeopathic OTC drugs to the same standards for other OTC drugs.[16] These actions may finally signal the end of homeopathy as the bedrock of your local drugstore's center aisles.

My guess is more studies will emerge to debunk homeopathy's mythological halo. In February of 2016, for instance, Professor Paul Glasziou, a leading British academic in evidence-based medicine at Bond University, declared homeopathy a "therapeutic dead-end" after a systematic review of 176 trials that focused on sixty-eight different health conditions concluded the controversial treatment was no more effective than placebo drugs. Writing in a blog for the *British Medical Journal,* Professor Glasziou stated, "I can well understand why Samuel Hahnemann—the founder of homeopathy—was dissatisfied with the state of 18th century medicine's practices, such as blood-letting and purging and tried to find a better alternative."[17]

It's one thing to use homeopathy for the placebo effect, but it's clearly another to put your health at risk if you reject or delay treatments backed by good evidence for safety and effectiveness.

"BUT MY ACUPUNCTURE AND HERBAL THERAPY WORKED"

So how do we explain the people who claim to have been cured by their unscientific methods? Can copper bracelets ease arthritis? Does chiropractic treatment help with ear infections and migraines? Will juicing stop cancer from spreading?

People who claim to have been cured or aided by their unconventional approach owe that success to the good ol' placebo effect, a remarkable phenomenon in which a fake treatment—an inactive substance such as distilled water, sugar, or a saline solution—can sometimes improve a patient's condition simply because the person expects or believes that it will be helpful. Contrary to how many people view the placebo effect, it's not casual mind over matter. Neither is it mind-body medicine. The *placebo effect* has become a catchall term for a positive change in health that cannot be credited to medication or treatment. This change can be due to many things, from spontaneous improvement and reduction of stress, to misdiagnosis, classical conditioning, or for reasons we have yet to understand scientifically. The placebo effect is an intense area of study today. We know that complex neuronal circuitry that can involve our emotions, hormones, neurotransmitters, and memory gives rise to placebo effects.

The placebo effect has existed since we first roamed the earth. In the eighteenth century physicians regularly administered inert pills when they had no real drug in their arsenal. By the latter half of the nineteenth century, the medical community began viewing disease in purely chemical and physical terms, so by 1900 placebos fell out of favor as therapy. That is, until the curious case of Mr. Wright sparked new interest in the phenomenon and inspired a new era of exploration of the science of the effect, which continues to this day.[18]

In the mid-1950s, a man who doctors called Mr. Wright was dying from lymphoma—cancer of the lymph nodes, which are an important part of the immune system. His neck, groin, chest, and abdomen were riddled with tumors the size of baseballs. The doctors had tried everything to no avail. Mind you, at this time medicine had little for treating cancer, unlike today, when powerful chemotherapies can dramatically improve the life and longevity of many cancer patients. According to a 1957 report by psychologist Bruno Klopfer, who worked at UCLA, Mr. Wright was optimistic that a new anticancer drug called Krebiozen would save him. He was already bedridden and struggling to breathe when he received his first injection of the drug. Three days later he was walking around and joking with the nurses. The tumors had shrunk by half, and after ten more days of treatment the hospital discharged him. But the other lymphoma patients in the hospital who also received Krebiozen showed no improvement.

That would have made a great, albeit somewhat unbelievable, story if it ended there. But over the next two months, Mr. Wright read press reports questioning the efficacy of Krebiozen. He grew concerned and then his cancer returned. What did his doctors do? They lied to him. They told him an improved, doubly effective version of the drug was arriving the next day. To say Mr. Wright was thrilled is an understatement. When the "medicine" arrived, the doctors gave him an injection; it did not contain any Krebiozen, but Mr. Wright vastly improved. He improved more so than the first time and walked out of the hospital symptom-free. Once again, however, reports about Krebiozen's uselessness emerged, and after two months of being healthy, Mr. Wright died within days of reading the bad news.

Mr. Wright's experience shows pointedly that a person's expectations and beliefs can mightily affect the course of an illness. But how do we explain this from a biological standpoint? What causes this to happen in a body? For it can't be just psychological factors tied to an inactive substance, can it? What's really going on?

Studies over the past several decades repeatedly show the efficacy of such sham treatments. Placebos can not only help alleviate ailments with a psychological component, such as pain, depression, and anxiety, but also lessen the symptoms of bona fide physical illnesses such as inflammatory disorders and Parkinson's disease. Occasionally, as in Mr. Wright's case, placebos have led to tumor shrinkage.

Some of the latest research has demonstrated that the placebo effect does not always come from a conscious belief in a drug. It can come from subconscious associations between recovery and the experience of being treated, from the feeling of getting a shot to a doctor's white coat and smell of an exam room. Which is how some people can experience the benefits of the placebo effect without necessarily believing in the treatment itself. That's right. Even if you *know* that you are receiving a placebo, also known as a sugar pill, you can still reap the benefits of the placebo effect. While this won't work for all illnesses, it can work for many illnesses that are symptom based, such as headaches, irritable bowel syndrome, or anxiety.[19] Such subliminal conditioning can control physiology, including the release of hormones and immune responses. Some of the most remarkable research to highlight the biology of the placebo effect has come from studying rats—animals that clearly cannot harbor beliefs about whether a certain drug will work.

Researchers have unraveled some of the biology of placebo responses, showing that they stem from active processes in the brain. One team of researchers, for example, conditioned rats by injecting them with the drug cyclosporine A—which suppresses the immune system and is used to prevent transplanted organs from being rejected by the body—while also feeding the rats water sweetened with the artificial sugar saccharin. The rats apparently associated the cyclosporine with the sweet drink, so that, later, feeding them the drink alone led their immune systems to partially shut down. The scientists hypothesized that the rats' brains sent messages to the immune system upon drinking the sweet water. Mind you, unlike humans, rats cannot consciously believe the drink is therapeutic, so some unconscious, associative learning causes these effects. Translation: a placebo effect does not hinge on a person's hoping for or believing in a positive outcome.

Plenty of research has also been performed on humans to help make sense of the brain biology and active brain processes that take place. For instance, in people with Parkinson's disease, use of placebos has been shown to increase dopamine binding in certain regions of the brain. The level of dopamine binding is correlated with the patient's perceived improvement in their outcomes. Other studies have further revealed that placebos have the power to increase neuronal firing in certain areas of the brain associated with Parkinson's disease and that these effects are correlated with improved motor performance. (Parkinson's disease is a to-date incurable progressive disorder of the nervous system that affects movement and can eventually render someone immobile.)

In psychiatry, the term *flight into health* is tossed around a lot and is similar to the placebo effect. Flight into health is said to occur when an individual seems to make a spontaneous recovery when faced with the prospect of therapy or of addressing particular issues. So, for example, people complaining of depression may suddenly pronounce themselves well upon leaving the therapist's office after just one visit. In some cases, the mere act of making the appointment will ease one's depression a bit.

We still have a lot to learn about the neurobiology of the placebo effect. No single pathway describes and defines the placebo effect in all scenarios. The placebo effect is not a definitive outcome in each case that's universal. It's fluid, ever changing, and dependent upon context—much like headaches. Headaches are all different, experienced differently in each individual, and are probably the outcome of unique biological pathways under disparate circumstances. But one thing is true: the placebo effect is real, so people such as Mesmer (who was unknowingly working with it) can't be all that bad if positive results are experienced in the absence of other authentic remedies. No doubt future research will eventually allow us to understand how a placebo-like effect can be leveraged for therapeutic purposes. It was just recently learned, in 2014, that the more invasive the procedure, the more significant the placebo effect—more so than medications.[20]

To quote Ted J. Kaptchuk and Franklin G. Miller in *The New England Journal of Medicine:*

> Medicine has used placebos as a methodologic tool to challenge, debunk, and discard ineffective and harmful treatments. But placebo effects are another story; they are not bogus. With proper controls for spontaneous remission and regression to the mean, placebo studies use placebos to elucidate and quantify the clinical, psychological, and biologic effects of immersion in a clinical environment. In other words, research on placebo effects can help explain mechanistically how clinicians can be therapeutic agents in the ways they relate to their patients in connection with, and separate from, providing effective treatment interventions. Of course, placebo effects are modest as compared with the impressive results achieved by lifesaving surgery and powerful, well-targeted medications. Yet we believe such effects are at the core of what makes medicine a healing profession.[21]

I have had firsthand experience with a type of "reverse" placebo effect. Or, maybe it was showing that medicine isn't always the answer, but a placebo certainly can be. Surgeons and anesthesiologists are always trying to find ways to minimize *emergence agitation,* which is the absolute misery and dysphoria that at least 50 percent of six-month- to six-year-olds experience after a brief surgery and anesthesia—even if the surgery is as quick as ninety seconds. Especially in infants, it's hard to tell if it's pain or just delirium. Usually it's a combination of both.

We recently did a study at UCLA to see if giving oral acetaminophen (Tylenol) twenty minutes before ear-tube surgery would reduce emergence agitation (assuming that the child was, in part, in pain from the ear surgery). We had three groups—a placebo group, where the child received a nonmedicated syrup; a low-dose Tylenol group; and a high-dose Tylenol group. Who fared the worst? With worse pain, more emergence agitation, and irritability? The high-dose Tylenol group! The low-dose Tylenol and placebo group fared about equally. The implications of these findings remain unclear. Perhaps the high-dose group had more stomach discomfort from more volume of medicine in their system; more likely, the higher level of agitation in the high-dose group was not due simply to chance—some children wake up more agitated than others, low-dose Tylenol, high-dose Tylenol, or no Tylenol at all.

I often also see placebo effects in the sidebar consults I get from family members of patients: "I'm doing a triathlon in seventeen days, and I think I have a sinus infection. Can you take a quick look?" They are already sitting in the exam chair of my office, their child on their lap, so I do what so many doctors swear they never will—I examine the family member, albeit cursorily. After a quick check, I often reply, "It's just a cold—rest a bit, drink a lot of fluids, and you'll be fine. You definitely don't need antibiotics." I swear I see an extra lift in their steps as they leave my office. Did my saying that they were not really sick make them better? Perhaps, in part, it did.

HYPE ALERT

* Complementary, alternative medicine works best in combination with conventional, traditional medicine.

* Homeopathic, "natural" medications do not undergo the same scrutiny by the FDA as conventional drugs.

* Homeopathic remedies can cause unpleasant side effects, interact with medications one is taking, trigger allergic reactions, and contain substances that may be either useless or, worse, harmful.

* The placebo effect is real and can sometimes work better than any drug. Complex neuronal circuitry that can involve our emotions, hormones, neurotransmitters, and memory gives rise to placebo effects.

10

TAKE A SHOT: THE PERILS OF LOSING THE HERD

How Vaccines Save the Community,
the Home, and Your Health

WHY IS THE ANTI-VAXX MOVEMENT SO STRONG TODAY?

WHAT IS THE CONCEPT OF HERD IMMUNITY?

DO VACCINES INTRODUCE TOXINS INTO THE BODY?

ARE SOME PEOPLE VULNERABLE TO SIDE EFFECTS OF VACCINES?

WHICH NEW VACCINE IS ANTICANCER?

One of my favorite *New Yorker* cartoons by Emily Flake features a boy sitting on an exam table in a doctor's office. The boy is riddled with red spots characteristic of measles. The caption reads, "If you connect the measles it spells out 'My parents are idiots.'" How true that statement is, especially today.

The word *vaccine* has garnered a lot of attention—and headlines—over the past few years. We are in some deep trenches of vaccine wars, and we can't get out.[1] Not too long ago, the startlingly low immunization rates in my city, especially in the most affluent Los Angeles communities, were made public. The pro-vaxx/anti-vaxx battles immediately heated up, punctuated by the measles outbreak that started at

Disneyland over the holidays in 2015. In 2017, Minnesota was struck by a measles outbreak, the state's largest outbreak in nearly thirty years.

Outbreaks of measles and other vaccine-preventable illnesses continue, in my city and others across the world. Living in this city as a parent of school-aged children, as a physician, and as a statistic, I have witnessed many responses to the 2015 and many subsequent outbreaks. Most of us have never seen a case of measles (or mumps or rubella or whopping cough or polio, to name a few), so we are scared that this microscopic monster will enter our neighborhood. Our school. Our house. My waiting room. On the flip side, some have a false sense that these contagions have vanished from existence or are certainly off the radar in their world of healthy living.

If you look at photos of kids in line for a polio vaccination in the 1950s, you see lively faces, silently—eagerly—waiting. When the photographer snaps the image of the needle piercing the skin of a child, the little one has not so much as a grimace. When I was a kid in the sixties and seventies, we certainly weren't as tough as my parents' generation, but getting shots was an unquestioned rite of passage—not a lifestyle choice. And the choice of when to get each shot was also not up for discussion—there was a schedule, which everyone who was fortunate enough to receive medical care followed. Now glance at any photo of vaccine administration in the last ten years or so. No waiting lines, at least not in this country. Especially not in my town.

Today's vaccine photos depict a radically different story. They invariably show a wailing, unhappy child and a cringing mother, holding her beloved in a near choke hold. This dichotomy reflects a historic societal shift in our feelings about vaccinations. No longer do we fear the viruses; we fear the shots that protect us from them. Moreover, regardless of what's in these needles, we now assume that children should not feel pain for any reason. Sadly, well-meaning parents threaten misbehaving children with a shot. Shots are not punishment and should never be portrayed as such. Yes, they hurt a bit. But so do skinning your knee, a muscle ache after a tough workout, or giving birth. But we all get back on our bikes, climb another mountain, and have more babies. Temporary pain is no reason to turn shots into an evil. But what about the real possibility of vaccine damage? The vaccine wars currently taking place are just as sincere as any blood-filled battle, so one side cannot be totally and utterly wrong, can it?

THE CASE FOR VACCINES

Unlike most any other health practice, vaccines are beneficial not just individually, but also for the greater good of society. Vaccines are (primarily) injected substances that are used worldwide to prevent communicable diseases. While they were initially created to fight viruses, such as the original smallpox vaccine, many current vaccines protect against viruses, such as influenza, measles, and rubella, and some prevent bacterial illnesses, such as pertussis (whooping cough), meningococcemia (meningitis), or diphtheria. Vaccines are not just for children. You are never too old to be immunized, and lots of modern vaccines are for adults and older folks. These include vaccines for shingles and certain strains of pneumonia. These are not illnesses you'd want to weather, if you even survive them.

The first vaccine was devised by using infinitesimal amounts of the virus (smallpox) to create an infinitesimal infection. As a result the vaccinated person would build up antibodies to this tiny exposure, later preventing contraction of the illness if exposed to a full-on viral infection. As vaccine science evolved, vaccines no longer needed to contain whole viruses (or bacteria) for a person to build up antibodies (or, become immune) to a given infection. Most vaccines nowadays contain only a small part of the virus or the bacteria, either part of the DNA, or part of a protein, to enable the recipient to form antibodies. The great benefit of this change in vaccine science has meant fewer and fewer side effects from vaccines. The downside is that some vaccines have become a bit weaker, meaning that more boosters are needed, and some people may not build enough antibodies from these weaker, albeit safer, vaccines. That's right—some people might receive all of the vaccines recommended, yet they are actually not immune to the illnesses the vaccines prevent. These folks have normal immune systems, but their antibodies just didn't kick in when vaccines were given. This is where herd immunity comes in. If most of the herd (i.e., population) receives vaccines, those who aren't individually immune will gain some protection from the herd's being immunized. If the herd (considered to be at least 90 to 95 percent of a population) isn't protected, infections start to seep in. We've seen outbreaks worldwide more and more in the past decade or two, mainly due to the double whammy of weaker, safer shots and weaker herds.

Vaccination is not solely a personal decision, nor is it even for one's family. Because of herd immunity, which I'll explain in more detail later in this chapter, being immunized impacts those around you in more ways than you can imagine. It shields not only you, your children, and your immediate family, but it also protects those around you—your neighbors, friends, and people in your community whether you know them or not. You choose to drive sober to save yourself, but also to save other drivers in your path. You choose (or are mandated) not to smoke in designated areas so as not to affect others and put their lives at risk.

What's also come to the forefront is delaying vaccines, or the trendy concept of creating your own schedule, based on the downright fallacy that this will somehow be gentler on a child. Even more fallacious is the idea that children with other ailments, such as history of premature birth, asthma, or allergies, are at higher risk for having reactions to the shots. In fact, the opposite is the case. Those kids who are more medically fragile need shots on schedule even more than the healthier ones. A child with asthma is at much higher risk of developing such complications as pneumonia from a vaccine-preventable illness such as influenza or pneumococcus than a child with healthy lungs.[2]

Vaccines are no more toxic to the body or a burden to the immune system than common exposures in daily living. Although some purport that vaccines inject contaminants into a child's pristine body, this is complete nonsense. Babies are exposed to billions of bacteria and viruses daily—beginning with their birth itself. Vaginal births expose a baby to millions of bacteria, both from the birth canal and the intestinal tract, before a baby lets out that first cry.[3] A kiss contains 80 million bacteria.[4] Breast milk contains hundreds of species of bacteria—not just from the mother's skin, but from the breast tissue itself.[5] Viral loads in vaccines are infinitesimally small compared to a day's exposure at a Mommy and Me class. And they don't overwhelm the immune system in a healthy human. The smallpox vaccine that my generation received presented a much greater challenge to the immune system than all the vaccines combined that kids today obtain over two years. Arguments about additives are often overblown or baseless. Take mercury, for example, which is a known neurotoxin and one of the most vilified additives in the anti-vaxx war.

Mercury exposure in felt-factory workers explained why Lewis Carroll's Mad Hatter was "mad." The expression *mad as a hatter* dates back to the eighteenth and nineteenth centuries, when hat-factory workers were exposed to high levels of mercury, which was used to stiffen the felt, leading to psychosis and dementia. In the 1950s and 1960s, Minamata Bay in Japan was exposed to high levels of industrial mercury pollution, leading to extremely high levels of mercury in the local fish. Newborns and children who consumed the fish regularly developed cerebral palsy, developmental delays, seizure disorders, blindness, and deafness. The form of mercury that causes central nervous system disorders in both adults and children is methylmercury.[6] The preservative thimerosal, which contains ethylmercury, has been thought to be the new culprit for mad hatter disease. While there is no association between ethylmercury exposure and central nervous system disorders, public health concerns about any form of mercury exposure were raised high enough to remove thimerosal from vaccines. But if thimerosal was removed from virtually all vaccines in 2002, and the rate of autism spectrum disorder diagnoses continues to skyrocket, how can one continue to associate thimerosal with autism?

To be sure, thimerosal is a mercury-based antiseptic and antifungal used as a preservative. Some flu vaccines do still contain thimerosal (0.025 mg/dose), but even flu shots containing preservatives are rarely, if ever, produced anymore. But consider where else you find much higher levels of the more damaging methylmercury:

* tuna (5.6 oz): 0.115 mg of methylmercury
* breast milk (1 liter): 0.015 mg methylmercury[7] (Any fish that a pregnant woman eats will have some mercury, which goes into her bloodstream and therefore the fetus. After delivery, maternal dietary mercury will also be in breast milk, albeit less so than in the bloodstream.)

Aluminum and formaldehyde in vaccines have also made headlines, scaring off too many for too many wrong reasons. Levels of these substances in vaccines are eclipsed by what we're exposed to in nature. Aluminum is the most common metal found in nature. It is naturally in the air, and in our food and drink. Infants get more aluminum through breast milk or formula than vaccines. Most of the aluminum that gets into your body is quickly eliminated. Formaldehyde is also ubiquitous in our environment and is produced in small amounts by most living organisms as part of normal metabolism. One pear contains sixty times more formaldehyde than a vaccine.[8]

Adults who avoid the flu vaccine and parents who host chicken pox lollipop parties have added even more twists and turns to the never-ending vaccine debate roller coaster. While many of us may have never had influenza, it's not an illness easily overcome without serious consequences to long-term health. This is true whether you're young or old. Every flu season, otherwise healthy people die from complications of the illness. The deaths include babies, kids, the elderly, and, yes, previously healthy adults with normal, robust immune systems. Those strong immune systems are what often lead to an individual's demise. Many healthy young adults have high levels of a substance in their immune system to fight off viruses. Ordinarily, this is a good thing, but when a bad infection such as the flu hits, this substance can act not only against the virus, but against the person as well, attacking lung tissue. This leads to complete respiratory shutdown and can be deadly. The body is caught off guard and responds by overreacting to the virus. So when you die of influenza, you essentially die from the damage your own immune system inflicts on your lungs.[9]

Although many adults today have had the chicken pox without any incident, this disease is not one to take lightly. Chicken pox infections have a significant risk of leading to pneumonia and even meningitis, a deadly neurologic disease. The vaccine, which may rarely cause a mild case of the chicken pox, is much safer than the illness.

A BRIEF HISTORY LESSON

Smallpox was a common epidemic of the eighteenth century and before. Remember Laura Ingalls Wilder's *Little House on the Prairie*? The family feared smallpox, and they were immunized in the first book of the series (otherwise, it might have been the first *and* last book of the series). Smallpox dates backs tens of thousands of years, when it would routinely decimate large populations throughout Africa, China, and Europe. It may have been responsible for the death of the Egyptian pharaoh Ramses V, whose mummified head reveals classic smallpox scars. Later, smallpox scourged Western Europe and landed in the United States along with European settlers. A century or so prior to a hint of the concept of vaccination, doctors in the seventeenth century found that giving a bit of fresh material, or pus, from a smallpox pustule to an uninfected person under their skin, via a sharp lancet, would provide some protection against the illness. This was termed inoculation, from the Latin *inoculare,* meaning "graft."

When the English aristocrat Lady Mary Wortley Montagu suffered from smallpox in the early 1700s, severely disfiguring her face and leading her brother to his death, she ordered that her children be inoculated. Word of the noble family's use of this technique spread, and inoculation gradually gained favor. The scientific concept of vaccination followed about half a century later. As is the case with many groundbreaking discoveries, the smallpox vaccine, the first and most powerful of all vaccines, happened from a chance observation. In the late 1700s, a small-town country doctor, Edward Jenner, noted that farmers and milkmaids exposed to cowpox never seemed to suffer from smallpox during its frequent outbreaks. They'd retain their beautiful complexions after a brief bout with the illness, unlike those who either died from smallpox or suffered mightily and had a scarred face to show for it.

Jenner began investigating if these workers were getting naturally vaccinated (*vacca* means "cow" in Latin) by exposure to the cowpox virus, which somehow provided protection against the smallpox virus. Cowpox was common among cattle at the time of Jenner's observations. It produced much milder symptoms than smallpox—its kissing, but more deadly, cousin. In 1796, Jenner encountered Sarah Nelms, a young dairymaid with cowpox lesions on her hands. He obtained material from these lesions and injected it into a young farmworker under his charge named James Phipps. This was before the days of informed consent, and certainly before the days of parental involvement in said consent. About a week later eight-year-old Phipps developed fevers, chills, some generalized discomfort, and loss of appetite. These symptoms soon subsided, and two months later Jenner injected Phipps with smallpox material. The boy stayed well, and Jenner concluded that his subject was protected from the deadly smallpox. His idea of vaccination, injecting a small amount of cowpox virus into healthy people, was not met with open arms. But in time, folks did hold out their arms to get the shot.

Those of us born before 1972 have the smallpox vaccine scar on our upper arm to show for it. Unlike the more modern vaccines, the smallpox vaccine carried such a high viral load, injected just below the skin's surface, that a local infection of smallpox would occur, followed by a telltale scar, sometimes up to one inch in diameter. After years of worldwide vaccination, in 1972 smallpox was declared eradicated in the United States; in 1977 a single case of smallpox occurred in Somalia for the last time, and in 1980 the World Health Organization considered smallpox to be eradicated worldwide.[10]

MODERN VACCINES

While the original vaccine, developed largely by happenchance about two hundred years ago, led to a minute version of the illness itself, this is no longer the case in today's world of infectious disease and immunology. Much of the vaccine avoidance among anti-vaxxers has stemmed from the idea that the only way to be protected from an illness is to contract a bit of the illness itself. Many people fear that vaccines will cause the illness against which they protect. I can't tell you how many adults I know will not get a flu shot because they think it causes the flu. While there is a molecule or two of truth to this, we have come a long way from the days of Laura Ingalls Wilder and her smallpox-infested clan. Today we have three types of vaccines: inactivated, live attenuated, and toxoids. The most common of these is inactivated; these vaccines do not contain viruses or bacteria, but simply parts of these organisms. These parts—either DNA, protein, or specific molecules on the virus's surface—enable your immune system to "see" this as a virus or a bacterium and gain advance notice. Your body then builds up antibodies to fight this germ, and these antibodies are present if your body is later exposed to the real thing. It's armed and ready long before the invasion. Often, these antibodies either don't hang around for your whole lifetime or aren't enough to protect you after just one shot, which is why booster immunizations are critical. Even individuals with a compromised immune system may receive these shots.

Live attenuated vaccines are most similar to the smallpox vaccine, but are given in infinitesimally smaller concentrations than the original smallpox vaccine. The majority of these confer lifelong immunity after one or two doses. People with compromised immune systems are unable to receive these. Live attenuated vaccines include the oral polio vaccine (now primarily administered as an injection, which is not a live virus, but an inactivated one), the measles vaccine, the rotavirus vaccine, and the yellow fever vaccine. A vaccine for tuberculosis, given in several countries around the world, known as BCG, is also a live bacterial vaccine and cannot be given to those with compromised immune systems. Another reason why herd immunity is so critical.[11]

The last form of vaccine is a toxoid, which is merely an inactivated bacterial toxin that an active bacteria would make. These vaccines include those against tetanus and diphtheria. A toxoid enables the body to create a defense against the bacteria if the real deal shows up. Small price to pay to stave off these pretty horrible illnesses. Most doctors who were born in the last hundred years have rarely, if ever, seen a case of tetanus. Fewer than thirty cases per year occur in the United States. Unlike communicable diseases, which travel from person to person, tetanus is a bacterial infection contracted through contaminated soil that enters an open wound. It can be fully prevented by a vaccine. When more common, the illness was known as lockjaw because it locks the jaw joints shut, and all body muscles become rigid and immovable. It leads to breathing problems, unstable blood pressure, heart problems, and death.[12]

Because of widespread vaccination diphtheria is also exceedingly rare in the United States, with only a handful of cases each year. But annually thousands around the world suffer from diphtheria. Unlike tetanus, this bacterium can be transmitted from person to person, leading to fevers, severe throat infections, airway blockage, and death.

Most of us have had virtually no exposure to these illnesses against which vaccines protect us, which is also increasingly contributing to vaccine resistance. Ignorance is, indeed, bliss. In my parents' generation, neighbors and friends routinely contracted polio. When the polio vaccine was developed in 1955 by Dr. Jonas Salk, followed by the oral polio vaccine by Dr. Albert Sabin in 1961, none from that generation in their right mind would consider not getting immunized against this horrible disease, and they certainly wouldn't hesitate to get their children protected.

During my childhood, chicken pox was a rite of passage. I had it in third grade, as one of the last cases in my class. Many parents still think that all kids should be exposed to the real illness and suffer through it for what they call natural immunity, rather than receive the vaccine. While, for the most part, chicken pox is relatively benign, *benign* usually means at least one week of itching misery, fever, and body aches for a child, school absences, and work loss for parents. As a medical student, we treated an eighteen-year-old who was hospitalized with chicken pox. He hadn't had it as a young kid, and he developed meningitis from the virus. A college friend of mine had chicken pox at age twenty-three, which knocked her down for two weeks, complicated by pneumonia. While these extreme cases are rare, even the seemingly benign ones are no fun. If I had to choose between one shot that hurts for a second or a week or two being home sick, I'd choose the shot.

Although the chicken pox vaccine has been around since the mid-1990s, many parents of millennials think the shot ruins a rite of passage. The folks who swear by natural immunity were up in arms, as if their child had been told that the tooth fairy wasn't real. The shot was taking away a special childhood experience. A trend, which likely continues today, followed of having a child with an active chicken pox infection lick a lollipop, and then sending it around to other homes. Parents would host chicken pox lollipop parties to expose all of the children present to the virus in a more "natural" exposure. The last lollipop would then be wrapped and mailed to another awaiting family, even across the country. No surprise: most, if not all, of the kids would be sick within the week. The idea that this natural immunity is somehow better than the immunization is nonsense. At worst, we need boosters to maintain protection. But we hardly need a week or more of illness that could be fraught with complications.

Another likely unrecognized benefit of the chicken pox vaccine is that it protects those who would have gotten to adulthood without ever having contracted the illness. Many of us remember the few kids who never had chicken pox. Lucky them? Quite the contrary, especially for girls. Those who get to their childbearing years who have never had chicken pox or been vaccinated are extremely vulnerable. Their fetus can suffer life-threatening complications if mom is exposed to chicken pox while pregnant. This can happen quite easily if, say, a toddler contracts the illness because he is not vaccinated and his mom, who also has never been exposed, becomes pregnant. The consequences can be devastating. Widespread immunization, however, prevents this.

I remember a medical school classmate of mine who was on her pediatrics rotation and became exposed to a child with chicken pox. She was terrified. None of us understood why—we were all over twenty-five, surely she had had chicken pox as a kid? But she hadn't, and because our age group missed the window of opportunity to get immunized in the 1990s, she wasn't protected. But the terror set in because not only was she not immunized, she was pregnant. If she contracted the disease, she was at risk of losing her baby and/or having a life-threatening illness herself. Thankfully, she did not become ill, and her baby was healthy. She did, however, get the vaccine soon after her baby was born.

THE PERILS OF PEDIATRICS

Countless magazine articles, television specials, and even feature documentaries have hashed out the vaccine controversy. Entire books have been dedicated to the vaccine controversy—many of them good, some of them great, and some of them just plain false. But all of these forms of media, no matter how valuable and valid, are words or talk. Only those of us who have experienced these illnesses that vaccines prevent or have treated them possess a deep understanding of the importance of vaccines. While I can't walk you back in time to the medical wards of a prior millennium, I'll try to set the stage for what went on.

When I was a medical student and in later years a resident, one of the drills (yes, like a fire drill) we would routinely create would be the treatment of a child with acute epiglottitis. The epiglottis is the leaf-like structure, made of cartilage, that sits at the top of the windpipe. It closes off the airway when you swallow, so that food doesn't "go down the wrong pipe." Otherwise, it's always open. If it swells from an infection, known as epiglottitis, you can suffocate. Epiglottitis is caused by the bacterium *Haemophilus influenzae*. There's now a vaccine for that.

The epiglottis is the tollbooth to the bridge. If the tollbooth is closed, you can't get on the bridge. The epiglottitis drills would go something like this: A child, usually around age four, would be brought to the emergency room in respiratory distress. He would be sitting up, drooling, and trying to cough. His face would be in a sniffing position—imagine leaning over a pot of soup, smelling the aroma. He couldn't lie down or he would stop breathing, so he keeps himself sitting up. Parents would report that he had had a mild cold, but woke in the middle of the night with a horrible cough, drooling, a look of panic, and a high fever. Here's how the drill would go: Don't touch the child! Upsetting the child by approaching him, touching him, or frightening him in any way might cause him to be more anxious, leading to crying, and more rapid and labored breathing, and subsequently the kid's airway would become completely blocked off as the epiglottis continued to swell. If you touch the child, he may die! If the child is kind enough not to be dying in front of your eyes, the next step would be to bring him with a parent to radiology—get one lateral X-ray of the neck. (We would be looking for a "thumbprint" sign in the upper airway—indicating swelling of the epiglottis. Radiologists call this a thumbprint sign because the swollen epiglottis looks like a thumb being held up, as if you're giving a thumbs-up sign. Ironically, this thumbs-up look is really thumbs-down for the patient, as it indicates the airway is dangerously blocked.) Bring the child, sitting up, with the parent to the operating room! Open up an intubation cart and a tracheotomy-tube tray. Keep the child on the parent's lap on the OR table. Try to give the child some oxygen and anesthesia via a mask as the panic-stricken parent watches. Remember: Don't touch the child!

If you can mask the child and he's able to breathe a little through the mask, but is getting sleepy from the anesthetic gas, have someone escort the parent to the waiting room. The anesthesiologist quickly establishes intravenous access. But then . . . you lose the ability to mask the airway and can no longer get air in beyond the swollen epiglottis. The heart monitor is trending up as the child's heart races. The oxygen monitor is trending down. Airway surgeons, be they musicians or not, can hear the difference between 100 percent oxygen saturation and 99 percent, but what you hear and see on the monitor is 80 percent and continuing down. The child's color turns from pink to blue to ashen. Then the worst happens—the heart rate slows, bradycardia, a sign of oxygen depletion and precursor to cardiac arrest. Per the drill and plan, you grab the tracheotomy knife, make a 2 cm vertical incision in the neck, continue through the second through fourth tracheal rings, spread the cartilage with a tracheal spreader, and pop a 4 mm endotracheal tube into that hole you made. Forty-five seconds.

Next time, do better. Not fast enough. Oxygen levels come up. Heart rate normalizes. Then you check your own pulse. Sinus tachycardia (fast, but normal, heart rate).

We dreaded epiglottitis drills. But these drills prepared us for the real thing. Unfortunately, drill as we did, the real thing rarely went as smoothly as how we practiced. Every year that nasty *Haemophilus influenzae* type B bacterium would usually affect three- to five-year-old children, and 6 percent of those affected would die. Drills or no drills. That same bacterium would routinely cause an eye infection called periorbital cellulitis. Kids would come in with one eye closed shut from redness and swelling of both eyelids. This bacteria was so contagious, we would treat the child as well as the rest of the family with strong antibiotics. And they would all be quarantined for at least a week. Sometimes this bacteria would cause bacterial meningitis, with permanent side effects including neurological deficits, deafness, or brain damage. None of this sounds familiar? The *Haemophilus influenzae* type B (Hib) vaccine became available in the 1980s and has been widely used since approximately 1999. Since then, I have never seen a case of epiglottitis. My residents and students have barely heard of it. It's history. Widespread use of this vaccine has reduced the incidence of Hib by 95 percent. While anti-vaxxers may still squeak by without this one, reasonable levels of herd immunity have kept Hib out of my operating room.

When I was six weeks old, my father was the camp doctor at a sleep-away camp in New England. That summer, a twelve-year-old camper contracted pertussis, better known as whooping cough. Even though she was quickly quarantined, I had likely already been exposed to the illness. It is caused by the bacterium *Bordetella pertussis,* and in its least aggressive form, it causes what's commonly called the hundred-day cough, as it triggers people to have coughing spells, or paroxysms, bringing them to their knees, to the point of passing out, for up to a hundred days. At worst, it causes these same coughing spells in infants, but infants often can't generate enough coughing power to expel the obstructing secretions from the illness, and they can suffocate and die.

Since my parents preferred that the latter not happen to me, my dad, a physician, got hold of a pertussis immunization (DTP, which stands for "diphtheria, tetanus, pertussis") and gave it to me as soon as he heard about this girl's diagnosis, even though it was two weeks before the recommended time for the shot, at age two months. And behold, I did not contract this potentially deadly, contagious disease. But escaping the illness had virtually nothing to do with my having been vaccinated on the spot (sorry, Dad, but thanks nonetheless). I was just lucky to have avoided exposure. For a child to be fully protected against pertussis, several vaccinations (in the form of boosters) need to be given, at two months, four months, and six months of age. I'm not sure why that girl got pertussis—whether she had not been immunized or perhaps she was not up-to-date with boosters. The DTP (currently known as DTaP, the *a* standing for "acellular") is an excellent vaccine, but, as with many immunizations, is not 100 percent effective and requires the three consecutive doses to confer immunity, and this immunity does not fully kick in until a few months after the final booster. DTaP is one of the safest vaccines, but has a relatively low efficacy, at about 60 to 70 percent. This means that a substantial percentage of people who are fully immunized do not necessarily make the appropriate antibodies to fight the illness. Here again is where herd immunity comes into play. If over 90 percent of a community is immunized, this will enable the entire group to be protected, even if some either don't respond to the vaccine or are not fully immunized. If fewer than 90 percent are protected, the illness can penetrate the herd, and epidemics can start.

A few years back, I was seeing an infant, about the age I had been during that New England summer, in my office. The baby had a piercing cough and seemed to be choking on her saliva, which she just couldn't clear. She was miserable. Her mom assured me the baby hadn't been like this until the past week. The day I saw her was on the heels of one of the largest pertussis outbreaks ever in Los Angeles County. Babies were dying. This baby had four older siblings, and the mom was a teacher. I examined the baby, and except for the horrible-sounding cough, her upper airway examination was normal. I asked the mom if her older children had been immunized, and she assured me that all four children were up-to-date. I expressed my concern for her newborn and suggested she go straight to her pediatrician's office, where they could do further testing and possibly admit the child to a hospital. I assured her that there was no surgical/airway problem, but it might be pneumonia or bronchitis. After she left the office, I called her pediatrician to let her know the baby and the mom were on their way, and I told her how relieved I was that all of the family members were immunized.

"Are you kidding me? That family has not seen one immunization in all of the years I've known them. Believe me, I've tried, but the parents have absolutely refused. She lied to you."

Lied? Okay, patients lie to doctors about how many cigarettes they smoke, how much alcohol they drink, whether they've stuck to their medication regimen, their sexual preferences, their diet, their exercise. But their children's immunization status?

"She probably knows that you could refuse to see her or get angry, so she gave the answer you wanted to hear," the clearly frustrated pediatrician went on. Apparently, she deals with the vaccine refusal regularly in her practice, and having been in medicine for over thirty years, having seen many children succumb to these now-preventable illnesses, her patience for these problematic parents was wearing thin. I realized then that the vaccine issue had become social, political, and divisive. Us against them. Them against us. We were no longer fighting the illnesses—we were fighting one another.

FOR THE GREATER GOOD

When my son was seven, he wanted to do a school project to raise money for kids in developing countries so that they could receive polio vaccinations. He thought it was cool that a shot cost a dollar, and that he could ideally raise enough money to protect about five hundred kids. He also thought it was neat that a kid's fingertip would be dyed purple after the immunization, as that was the only feasible method of record keeping in such rural communities. All of this was simple math. The complicated part was explaining to him that, while he wanted to strengthen a herd in need across the globe, the herd in his own back-yard was crumbling. How would I explain to him that this seemingly lifesaving project was actually a hot-button issue that would offend up to 20 percent of his school community, so much so that the project was tabled because it could offend? I tried to explain, but he didn't get it. Neither did I.

Vaccine refusal is not new. It did not originate on Google, in Holly-wood, or on a Twitter feed. Vaccine refusal is almost as old as vaccines themselves. The pioneer of vaccines, Edward Jenner, was hard-pressed to convince prairie pioneers of the time that receiving a small amount of the cowpox virus would protect them from the more deadly small-pox illness. They were (rightly) terrified that even an infinitesimal amount of viral exposure, even for the purpose of protection, would cause the dreaded disease. In 1855, the State of Massachusetts began requiring vaccines for schoolchildren. England was tougher. In 1853, the Vaccination Act made compulsory the immunization of all babies against smallpox unless they were considered medically unfit. So began the first medical exemption for vaccines. Objections to this act led to the earliest anti-vaxx movement, a good 150 years before social media and even the theory of autism linked to vaccines. A clause to the Vaccination Act enacted in 1898 added a "conscientious objector" loophole, enabling people to opt out. The first "personal belief exemp-tion," long before debates in the commonly known pockets of entitle-ment and juice bars throughout this country, was more than a hundred years ago. Even back in nineteenth-century England, approximately two hundred thousand conscientious exemptions to vaccination were granted in the first year of this clause's enactment alone.

In Barry Glassner's 1999 seminal book, *The Culture of Fear*, the sociologist describes the nationwide panic after a television special that aired in 1982, entitled "DPT: Vaccine Roulette." The show depicted anecdotal, albeit heart-wrenching, stories of children who became severely handicapped or even died, after receiving the DPT vaccine. Needless to say, this piece was picked up by other national media outlets, and the panic set in. Despite that physicians and the FDA presented large long-term studies with a combined sample of over 1 million children, and data dating back to the introduction of the DPT vaccine in 1949, showing that in years prior as many as 265,000 children annually came down with pertussis and 7,500 died, the spread of vaccine fear was already rampant and became viral long before there was You-Tube. What followed was a powerful group with the same acronym, Dissatisfied Parents Together (DPT), led by Barbara Loe Fisher, who claimed that her son developed acute neurological devastation from the DPT vaccine. The media jumped on this bandwagon, and within two years, several of the DPT manufacturers fell out of the market, and there soon was a vaccine shortage.

This was not unique to this country. Ten years prior, a vaccine scare in Japan led to the banning of this vaccine, spurring a tenfold increase in whooping cough cases and tripling of whooping-cough-related deaths. Panic in England led to a 40 percent decline in immunization, even though the vaccine was available, and a resultant surge of one hundred thousand cases of whooping cough over the following years. To stave off mass hysteria, endless lawsuits, and anecdotal events from making headlines and leading to widespread panic, Congress financed the Vaccine Injury Compensation Program, a no-fault plan that would compensate parents who believed their child had a vaccine-related injury. This would also prevent a nationwide (or even worldwide) public health crisis and free the courts from interminable lawsuits from vaccine-related adverse events. But cultural anxiety and media-savvy advocates will keep the vaccine fears coming, despite all the data on the planet, all the legal protection for pharmaceutical companies, and all of the evidence of the safety and efficacy of vaccines.

For some vaccine refusers, it is simply a control issue: I know what's best for me and my family—a mother's (or a father's) instinct. We'll choose to immunize (or not) if and when we feel it is right for our family. Dr. Bob Sears, who practices just down the freeway from me, has been a proponent of an alternative vaccine schedule. Parents across the nation have taken to "Dr. Bob's" schedule.[13] These kids are not protected against measles, rubella, chicken pox, hepatitis A or hepatitis B on the schedule recommended by the CDC and AAP.[14] Dr. Bob even claims, "My schedule doesn't have any research behind it. No one has ever studied a big group of kids using my schedule to determine if it's safe or if it has any benefits."[15]

In fact, studies in 2010 and 2013 showed no difference in developmental outcomes of children who received vaccines on the recommended schedule compared to those who used a delayed approach.[16] For some, it's a conspiracy issue: the pharmaceutical companies, medical groups, and government are out to get us. For others, it's an issue of ignorance: Those illnesses can't be *that* bad, or if they don't exist anymore, why do we need shots? For many, it's a fallacious idea of chemicals and so-called toxins—too much too soon will harm my baby's body. And aren't pharmaceutical companies, scientists, corporations, and doctors in it for the money? This last issue has fully entered the vaccine debate. I can speak firsthand for doctors and say that we barely break even financially giving immunizations. More often than not, they cause a slight financial loss. Physicians' offices buy vaccines, buy the syringes and needles, pay the nursing staff, and pay the rent. Immunization costs are not hugely marked up, so believe it or not, vaccinations are one of the last vestiges of pure medical care: to prevent illness. I don't know any physicians who like to see a case of measles or mumps. Vaccination is one of the few absolute non-moneymakers in our profession. Now, giving a Botox injection—that's another story.

Before a vaccine is considered safe for use in people, it undergoes some of the most extensive testing in medicine. A typical vaccine undergoes up to fifteen years of rigorous research in safety and efficacy studies before it is brought to the public. This is substantially more than for most any other over-the-counter or prescription medicines, and certainly more than any homeopathic remedy, which often undergoes zero years of research and zilch evaluation by an outside agency such as the FDA before landing on the shelves at your natural foods store. Despite this, many continue to feel that the FDA is not being up-front with the public, and that the government is hiding dangers and ulterior motives. A widely disseminated conspiracy theory on the vaccine front is that the FDA, Big Pharma, and even pediatricians are all in cahoots to make a big profit on vaccines, often taking shortcuts on safety checks in the name of dollar signs. This couldn't be further from the truth. Vaccine production and administration has one of the lowest profit margins in pharmaceuticals.[17] In the past years, in efforts to quell these concerns, the Institute of Medicine, now called the National Academy of Medicine, has acted as a further level of protection and review, ensuring that all study data reviewed by the FDA become available to the public. The Health and Medicine Division of the National Academies of Sciences, Engineering, and Medicine publishes this information annually.[18]

THEN WHAT EXPLAINS
VACCINE-RELATED DAMAGE?

Despite so many levels of safety regulation and confirmation, vaccine-related injuries do occur. Most vaccine reactions are not actual injuries and are mild and short-lived. Usually, some pain occurs at the injection site, or even some swelling. Some get a fever, muscle aches, or a local skin rash. These reactions may occur within the twenty-four hours of a vaccination, or weeks later. Rashes can look like measles or chicken pox. The fifth DTaP booster, given at age five years, hurts and may cause a painful large bump or redness at the injection site.

But those minor reactions are not what prevents people from getting themselves and their kids immunized. It's the big stuff, both real and imagined. Yes, there can be high fevers, and even grand mal seizures, after a vaccination. But, no, vaccines do not cause autism. Period. I've heard and read stories about vaccines causing SIDS (sudden infant death syndrome), multiple sclerosis, asthma, and type 1 diabetes. This is again what one might call "true, true, and unrelated." Remember, vaccines are given in such high volume in the first years of life, and these diseases do exist, many of which, especially SIDS, asthma, autism, and even type 1 diabetes, present in the first months or years of life. Autism is often diagnosed sometime between ages one and two, around the time that the MMR vaccine is given. The diagnosis of autism around the time this vaccine is received is coincidence. That is all. I've heard of parents refusing or delaying vaccines for their children because of the belief that vaccines cause the illnesses against which they protect us, and that they cause autism. The autism conjecture stemmed from Andrew Wakefield's article in *The Lancet* published in 1998. As I've already detailed, this article was later found to be fraudulent and was pulled from the electronic records of this journal.

Vaccines do, however, cause real and horrific adverse reactions. Vaccine-related deaths can occur and are impossible to predict. Vaccines can cause encephalitis (inflammation of the brain), and severe anaphylaxis (life-threatening allergic reactions). The VAERS (Vaccine Adverse Event Reporting System) keeps records of these events. But the numbers tell the real story. Since 1988, approximately fifteen thousand claims have been brought to this compensation plan. Let's assume that all of these claimants were justified in thinking that their child's adverse event was due to vaccination. Here are some other figures: About 4 million babies are born each year. In the first year of life, a baby receives anywhere from eighteen to twenty-two individual immunization shots, many of which contain up to three vaccines (for instance, DTaP contains three vaccines—diphtheria, tetanus, and pertussis). In one year, babies from ages zero to twelve months receive 80 million shots. Children between ages twelve and twenty-four months are given another 24 million shots. Another 16 million shots are given to two- to ten-year-olds. And that doesn't include the annual flu vaccine. So from 1988 to 2016, each group of children born each year ages zero to ten years receive 120 million shots. In total, children born between 1988 and 2006 have received no less than 3.36 billion shots. There have been fifteen thousand claims. Assuming all claims were legit, that's .0000004 percent of shots given had some reportable adverse event.[19]

WARTS AND CANCER

One particularly busy clinic day, the on-call resident called me urgently. A four-year-old who had been diagnosed with asthma and had for the past two years had breathing treatments—hospitalizations to receive steroids—and was advised not to play or run with his friends was in acute respiratory distress with noisy breathing. A flexible-scope exam showed large masses of fleshy tissue filling his entire voice box. He had no voice and could hardly breathe. We rushed him to the OR. If he had a voice, he'd have been crying. Certainly his parents were. He was trying to die. With my anesthesia team, we got an intravenous going and had him breathe with whatever tiny airway he had through a mask to get him off to sleep.

When I took a look with a rigid laryngoscope and telescope, his airway was a mess of pink masses—no identifiable structures. I squeezed a tiny endotracheal tube through the bloody mess, covered his face with wet towels, which is a standard precaution during all laser airway surgeries to prevent a laser fire, and got the laser going to obliterate the masses. Within an hour or two, we saw some semblance of vocal cords and got him safely to the recovery room, into his parents' arms. He was hoarse, but he could breathe. Sounds great, right?

Well, he came back two weeks later, and two weeks after that, for the same procedure. The intervals spread out over the coming years, but he will never recover from this disease. To any airway doctor, the diagnosis was easy, albeit devastating: respiratory papillomatosis—a chronic disease of masses of the airway, seen most commonly in young children, with many treatments but no cure. Approximately twenty-five hundred new cases per year are diagnosed, and while treatments have come a long way, many children continue to need four to ten surgeries per year for most of their childhoods, and into adulthood as well. Many require placement of a tracheotomy tube to breathe, and many die each year. This child, now an adult, continues to have this disease.

He became my patient in 1999. Why would I be bringing this up in a chapter about vaccines? His illness is caused by HPV, or the human papillomavirus, a widely prevalent virus, present in the genital tract in up to 80 percent of the U.S. adult population. It is spread via genital-genital contact, oral-genital contact, genital-anal contact, oral-anal contact, and oral-oral contact. Needless to say, it spreads far and wide. Most have heard of genital warts and anal warts (these are due to HPV). But few know that HPV can lead to cervical cancer, and even fewer know that HPV leads to obstructing papillomas of the airway. While a high percentage of childbearing females carry HPV in their genital tract (even without having visible warts), an extremely small percentage (a total of about three thousand per year) transmit the virus in utero to their unborn child. The virus then multiplies in the child's respiratory tract, until, typically in the toddler years, obstructing warts form and block the child's airway. It remains unclear which women pass this virus along to their children. While many used to believe that cervical and vaginal warts were the direct cause, cesarean section was later found not to protect against spreading this disease and that the transmission was in utero. The good news: had my patient been born more recently, his condition could have been prevented because now there is a vaccine against this virus. Most commonly known as Gardasil, the vaccine has met with so much controversy that its short-term benefits (fewer genital warts in the population) and long-term benefits (fewer, or eradicated, respiratory papillomas, and fewer cases of cervical cancer) cannot yet be assessed, as the percent of the target population actually receiving this vaccine remains quite low.

A half million new cases of cervical cancer alone are diagnosed per year around the world, with over 250,000 deaths.[20] In the United States, it has a price tag of $350 million annually, and it's often due to HPV. Oral cancer treatment tacks on another $38 million, anogenital warts cost $220 million, and airway papillomas cost $151 million. Michael Douglas's throat cancer, which he called "oral sex cancer," was caused by HPV. A large percentage of oral cancers, especially tonsil cancer, are HPV in origin, so his was not an unusual case. While the vaccine costs about $1.7 billion to administer, the total cost of HPV-related illnesses is over $6.5 billion annually. But it's not cost that's giving people pause.

The anti-vaxx community is also opposed to this vaccine—not because they think it causes autism, but because it might cause cancer or other debilitating side effects. Claims have also been made that giving girls this vaccine will somehow give them the sense that they are now free to have sex, and that the vaccine will promote early sexual activity in adolescents. This has been debunked.[21] And the claims that the vaccine causes cancer are nonsensical. It doesn't cause cancer—it *prevents* it, and so many other horrific conditions. But because it's not yet widely used (only 20 to 40 percent of those eligible actually receive it), the benefits cannot be assessed. The herd immunity needed is likely around 80 percent, and it would take at least one or more solidly immunized generations to start to see benefits to offspring. But because we're not even close to reaching this, we still don't know.

Indeed, some small groups of adults and children cannot and should not receive vaccines. People with compromised immune systems, either from an illness or from medical treatments such as chemotherapy, cannot receive many of the vaccines. If patients have had an illness, such as chicken pox, they don't need to be vaccinated against it (although no harm would be done if they were). If patients have a known allergy to one of the vaccine's contents, such as egg protein, they may not be able to receive certain batches of the vaccine. However, even these vaccines have batches with their contents altered so those with certain allergies can receive them.

It is often said that it takes a village to raise a child. Well, it takes a herd to protect and grow a village. The concept of herd immunity has made it beyond the immunology lab to the sphere of social media and even Penn & Teller's comedy routine in recent years. The concept is quite complicated, as there is no magic percentage to confer herd immunity for all vaccine-preventable illnesses for every population. However, the concept of herd immunity is that if a critical mass, or percent, of a population is protected, i.e., immunized, against a certain illness, the nonprotected individuals will gain some protection by the herd's being protected. For most immunizations this is in the 90 to 95 percent range. My family and I are all immunized. You're welcome.

HYPE ALERT

* Vaccines are no more toxic to the body and a burden to the immune system than common exposures in daily living. You are exposed to more aluminum and formaldehyde, for example, in nature through air, food, and drink than what you'll get in a vaccine.

* The smallpox vaccine that my generation received presented a much greater challenge to the immune system than all the vaccines combined that kids now obtain over two years. What's more, arguments about additives are often overblown or baseless.

* Pharmaceutical companies, scientists, corporations, and doctors are not in the vaccine business for the money. They are in it to save lives. Vaccine production and administration has one of the lowest profit margins in pharmaceuticals.

* Vaccines do not cause autism. Vaccine-related damage can occur in some people, but the chances are exceedingly rare—much rarer than those of contracting the illnesses that the vaccines prevent.

* Vaccines against the human papillomavirus prevent some forms of cancer, such as cervical and throat.

* You are never too old to be immunized, and lots of modern vaccines are in fact for adults and older folks.

11

TESTING, TESTING, ONE, TWO, THREE: FROM THE OUTSIDE LOOKING IN

How to Determine When to Get Checked,
X-rayed, Swabbed, or Poked

DO MAMMOGRAMS AND COLONOSCOPIES REDUCE THE
LIKELIHOOD OF DYING FROM BREAST AND
COLON CANCER RESPECTIVELY, OR IMPROVE OUTCOMES?

DO FALSE POSITIVES ON PSA TESTS (PROSTATE SCREENINGS)
LEAD TO UNNECESSARY, RISKY PROCEDURES?

ARE FULL-BODY SCANS HELPFUL OR HURTFUL?

Every day, we each make an average of thirty-five thousand remotely conscious decisions.[1] That's about a decision every two and a half seconds. (In contrast, young children only make about three thousand decisions each day.) That's a lot of judgment calls, though the majority of them happen without our thinking about them. Driving a car alone entails multiple decisions every second. When it comes to mulling health-related options, however, I find that even the most health-conscious, educated consumers can be confused by life's simplest choices. This is largely due to the volume of conflicting information in the media.

Now, could you answer the above questions definitively and with confidence? Don't panic or feel bad if the answer is no (and don't google). All of these questions have surprising answers and implications about the pros and cons of medical testing today. Indeed, some of these tests and their value have been hyped. Patients are often not given implications of false negatives or false positives, and how tests such as body scans, while revealing, can uncover what are called incidentalomas—tumors found by accident that can lead to further unnecessary testing and even surgery. These tumors might never cause disease. Sometimes finding them can save your life, but sometimes finding them can end it. The same holds true for DNA tests: they can spot mutations that are associated with serious illnesses, but that doesn't mean you will ever develop those illnesses. Such tests detect potential risks, not illnesses. So the question becomes, Do you want to know if you carry gene X if it's linked to disease Y? How much information is too much? When does the information cause harm? Addressing this topic affords me the opportunity to share the nuances of medicine and health care, as well as the importance of tempering data with personal values and even probability. Patients must be treated case by case; by no means does one size fit all. Ethics also come into play, and hard choices are increasingly being made in modern medicine now that we have technologies to identify genetic forces. For example, consider the real-life situation I mentioned early in the book—my colleague whose husband has Huntington's disease.

In 1872, George Huntington studied several generations of a family living in East Hampton, Long Island, with similar neurodegenerative symptoms, including mood and movement disorders, progressing to severe neurological devastation and early demise. Before the advent of Mendelian genetics, which governs our current understanding of the laws of inheritance, Huntington identified the concept of autosomal dominant inheritance.[2] A person with Huntington's disease has not only a 50 percent chance of passing along the gene for it to children, but also a 50 percent chance the children will have the disease. The gene for Huntington's is autosomal dominant, meaning that a person only needs one copy of the gene for it to express the disease. Most inherited traits are either autosomal recessive, meaning you need two copies of the gene to express the trait, or sex linked, meaning that a male needs one copy to have the trait, but a female needs two, as a sex-linked trait is carried on the X chromosome (males are XY, and females are XX). A disorder that's autosomal dominant, such as Huntington's, carries much more weight in the genetic pool, as even if one parent has it, children have a 50 percent chance of having it. Other familial disorders usually drop to a 25 percent chance when autosomal recessive.

While autosomal dominant inheritance patterns had been recognized for over a century, the gene locus, on chromosome 4, was not found until 1993. The location for the gene for Huntington's disease was the first chromosomal gene locus to be discovered. The family situation of my colleague whose husband has Huntington's has been challenging, although not unique among the many families with this disease. Her husband decided to get tested for Huntington's soon after his own mother was diagnosed. Although he had not yet developed any signs or symptoms, he just wanted to know. When the results came back positive, he and his wife broke down in tears—not just because of the news of his diagnosis, but because she was pregnant with their first child. Despite knowledge of the ticking time bomb—symptoms were bound to kick in at some point—life went on as usual. He had no problems, and the couple soon had three healthy children. When the oldest child was ten, the husband started to show signs of the disease—not the well-described issues with motor function, but anxiety and mood swings. This is a common, yet not-well-publicized, presentation of this illness. As tensions grew in the household, my colleague decided to tell the kids. She said she'd prefer them to know that daddy had an illness they were trying to treat, as opposed to their thinking that he was just moody and mean. She still remarks on how resilient her kids are—how they handle their father's illness with grace and empathy. Do they know they each have a 50 percent chance of not only carrying the gene, but also having the disease itself? Yes, as they are all now approaching or are in their teenage years, they know. Do they want to know if they carry this gene? For now, no. And their parents are respecting their decisions to wait. Right now, they are focused on their dad. In the future, they will make decisions about where to go to college, what to major in, and if and when to marry. And they will each decide whether to be tested for Huntington's. Perhaps by then better treatments will be available.

I am still overwhelmed with awe whenever I discuss this with my colleague. Her and her family's perspective on big decisions regarding health diagnostics puts genetic testing, or any other testing, on a whole new level. These are real numbers, and the numbers are big, and not in anyone's favor. We're not talking about being at risk for an illness, and one with an array of treatment options. We're talking about a disease that currently has no cure, and no good long-term outcomes.

I treat many families who have a child with congenital hearing loss, meaning the child was born with the condition. As I described early on in the book, we have now located a gene to identify the cause of many types of hearing problems. If we test a baby for this gene, should we test the normal-hearing siblings, who may be carriers? Is hearing loss to be considered a true deficit, or a trait variation that many embrace? If the siblings are under eighteen, do they have a choice to get tested or to wait?

No single solution exists for any of these tough questions. And one can fall prey to hype in the alleged benefits (and, conversely, pitfalls) of these tests. But with the information I provide in this chapter, I trust you will feel empowered and enlightened.

FIRST, DO NO HARM

A good question: How do we calculate and determine a treatment's potential to do harm to people? After all, not everyone who takes a certain medication, for example, will experience a negative side effect, and not everyone who takes the medication will benefit from it. To answer this question, we have to look at what's called the NNT, which stands for "number needed to treat." Conversely, we have the NNH, or "number needed to harm." The NNT tells us how many people would need to receive a medical therapy for a single person to benefit. A simple example of this, as well described in a *New York Times* article by Aaron E.

Carroll and Austin Frakt, is aspirin. The NNT for aspirin to prevent one additional heart attack over two years is two thousand.[3] In other words, for every two thousand people who take aspirin, one will avoid a heart attack. Given aspirin's touted benefits in addition to helping reduce risk of heart attack (e.g., anti-inflammatory and, in turn, anticancer—see below), you might think it's worth taking the risk of its not doing anything for you. But you then need to consider the flip side: aspirin's ability to do harm in some people, perhaps you.

Aspirin's potential side effects include an increased chance of spontaneous bleeding in the stomach or head. Life-threatening bleeding can also occur from a minor head injury, turning a little bump into a bloodbath. If you're taking aspirin and have an undiagnosed small gastric ulcer, aspirin can trigger some pretty serious intestinal bleeding. Granted, not everyone who takes aspirin bleeds. And some people will bleed whether they take aspirin or not. So how do you decide whether to take aspirin? Well, a low-dose, daily baby aspirin (81 mg) may not only impact heart disease but may also prevent cancer. In a seminal study published in *The Lancet* in 2011 that reviewed more than twenty-five thousand patients, taking daily aspirin for five years or more significantly reduced the twenty-year risk of developing colorectal cancer, lung cancer, brain cancer, and pancreatic cancer.[4] The current thinking is that aspirin has anticancer properties due, in part, to its ability to reduce inflammation and tumor cell growth by interrupting tumor cell signaling. This was only found if taken for five years, and the primary reduction was seen in patients over sixty-five.

Another systematic review of multiple studies also showed benefits from taking aspirin long term. As did other reviewers, the authors found that aspirin had to be taken daily for at least three to five years to be associated with lowered cancer risk. The risk of bleeding increased both with increased aspirin dosage (and higher dosing did not equal better cancer protection) and advancing age. Fifty- to sixty-five-year-olds who took daily aspirin (70–325 mg/day, roughly equivalent to one baby aspirin to one regular-strength adult aspirin) for ten years saw a relative reduction of between 7 percent (in women) and 9 percent (in men) in several types of cancers, myocardial infarctions (heart attacks), and strokes over fifteen years. The overall death rate dropped by 4 percent over twenty years. The authors found an overall favorable benefit-harm profile for low-dose, long-term daily aspirin use in older adults. But they also acknowledge that there are known risks, and that individuals who are at increased risk for bleeding need to be better identified for future reduction in risk profiles.[5]

But here's how the headline might read: "Aspirin Prevents Cancer!" Without the caveats of daily use, only specific cancers, and only in older patients. And nothing about associated risks of aspirin use, including life-threatening bleeding. I fit the criteria for aspirin use to prevent cancer. I'm also worried about bleeding, even though I have no known risks (except for my early-morning runs in the dark, when I may trip and crack my head open). Given this risk, I do what any statistically minded person who knows that good data could flip to bad at any time would do: I take one baby aspirin every *other* day.

Now let's consider mammography, which screens women for breast cancer. According to the American Cancer Society's guidelines as of 2016, women ages forty to forty-four can consider mammography; women ages forty-five to fifty-four should have annual mammograms; and women ages fifty-five and older can consider continuing with annual mammograms or switch to mammograms every other year. These recommendations go along with self- and health-care-provider breast examinations and variations for personal and family history of breast cancer and any other predisposing risks such as known genetic risk factors for cancers.[6]

But you might be surprised by mammography's NNT, based on data from randomized controlled trials of breast cancer screening: 1,477. Translation: to prevent one death from breast cancer after thirteen years of follow-up, you'd need to conduct mammography on 1,477 women. And that one woman would likely have died from something else entirely, such as another type of cancer or a heart attack. So in this scenario, mammography provides no benefit at all in terms of preventing death from all potential causes. On the flip side, mammography does have its downsides (NNH). Mammograms can entail false positives and overdiagnoses. The ensuing treatments such as chemotherapy, radiation, and/or surgery provide zero benefits but do come with risks of harm. Would you want to undergo treatment for cancer if you didn't have cancer?

What's more, in February 2015, an alarming study emerged in *The Journal of the American Medical Association* showing that biopsy specialists often misdiagnose breast tissue.[7] Such inaccurate diagnoses lead to overly aggressive treatment for some women and undertreatment for others. Evidently, pathologists are adept at determining when invasive cancer is present in breast tissue, but are less precise at calling the right diagnosis with less severe conditions or when biopsied tissue is normal.

Let's bring some numbers into this conversation. The data shows that for about every fifteen hundred women who get mammograms for ten years, one of those women might be saved from dying of breast cancer. But here's the rub: She won't be saved from dying from something else. And about five women in the mammography group would undergo surgery and about four would receive radiation—treatments with adverse side effects. Once again, how can a woman weigh these pros and cons—especially as they pertain to her own unique body and personal risks—and have a thoughtful conversation with her doctor about the relative potential benefits and harms of treatments?

Although we doctors hope that every treatment offers a benefit, in truth the benefits are often much less likely and impactful than we anticipate. I should add that these numbers—the NNH and NNT—are merely statistical concepts used in research and scientific circles; you are not going to find them posted in your doctor's office or listed on your prescriptions. (To learn more about these numbers, go to www .thennt.com, where you'll find a library of data based on clinical trials.) Despite all the data, be it positive, negative, or mixed from such a critical screening tool as mammography, you, your mother, your wife, your sister, or your friend are not fifteen hundred women—but just one woman.

While we rely on data from large studies to help make decisions for both screening and treatment, when you're talking about yourself, a family member, or a friend, the NNH is one. The good news is that these decisions are not made by one person. One doesn't (or at least one shouldn't be able to) just show up at mammograms-are-us to get one's boobs squeezed like a panini on a grill. So much goes into individual decisions to screen: age, family history, pregnancy history (pregnant and nursing moms often hold off on mammograms), as well as discussion of other potential screening tools such as breast physical exams, breast ultrasound, breast MRI scans, and 3-D mammograms. All of these have pros and cons, and no matter what the largest studies show, the breasts on that one patient are the only ones that matter. This is why deciding when to do mammography is determined with your own physician who knows you and your risk factors.

THE FORGOTTEN MERITS OF H&P

Have you ever left a doctor's office with the feeling of "They didn't really do anything"? If no X-rays were performed, no blood tests were ordered, and no medications were prescribed, you might wonder if your visit had any value. (And that darned surgeon didn't even recommend surgery?!) Believe it or not, listening to patients, and examining them *without* high-tech tests, can sometimes be more valuable than the most up-to-date fine-cut imaging body scanner. The history and physical examinations are the oldest, and in some ways remain the most accurate, diagnostic tests. But because they are low-tech, many will disregard them as inconsequential. Doctors have been among the first to throw this art of patient care out the window. When a doctor sees a patient nowadays, an average of no more than fifteen seconds go by before the doctor interrupts the patient's story. The insane pressures to see higher numbers of patients in a shorter and shorter time, combined with our increasing reliance on laboratory and radiological tests, often leaves little or no room for good old listening, asking the right questions, and performing a careful physical examination with experienced hands, ears, and eyes. But this is all a doctor needs. Whenever I can, I remind myself of this and, with great effort, shut myself up and just listen. It's always amazing what I learn. But for the sake of efficiency, we often order diagnostic tests in lieu of human contact, sometimes even before laying eyes on the patient.

A perfect example of this sea change of diagnostic testing superseding the classic history and physical (H&P) is the diagnosis of one of the most common problems seen in emergency rooms: acute abdominal pain. An ER doctor's first order of business is to rule out appendicitis, or inflammation of this little vestigial organ at the junction of your small and large intestines. Over twenty-five years ago, I completed my general surgical training with one of the masters of the written H&P: Dr. William Silen, at Boston's Beth Israel Hospital (now the Beth Israel Deaconess Medical Center). His own surgical training, and the vast majority of his years of clinical practice, took place in the days before CT scans, MRIs, portable ultrasounds, and many of the laboratory tests we routinely use today, often before even meeting a patient. Worldwide, Dr. Silen is known for his published updates and revisions of the text *Cope's Early Diagnosis of the Acute Abdomen,* first written by Dr. Zachary Cope in 1921, six years before the birth of Dr. Silen. The twenty-second edition was published in 2010 by Dr. Silen, and he has been responsible for each edition since 1980, just a few years after Cope's death.[8]

Since Cope's first edition nearly a century ago, the anatomy and pathophysiology of the human abdomen has not significantly changed, but its evaluation and management have shifted astronomically. It can be exposed to new drugs and to new foods responsible for acute gastroenteritis (what you'd call food poisoning), and the list of differential diagnoses for acute abdominal pain has evolved ever so slightly. But these changes pale in comparison to changes in the available diagnostic testing on one of the most common complaints in any emergency room across the world.

Dr. Silen was a larger-than-life surgeon and consummate doctor and had many sayings that stick with me and so many of his former trainees even today. He was the master of listening and perfected the art of the H&P. He was famous for teaching, "If you listen long enough, the patient will tell you what's wrong with them." When we were stumped in finding an answer while making hospital rounds, he'd state, "Dial 1-800-HISTORY." On countless occasions, he'd walk into a patient's room, ask the patient a question that seemed so obvious, but one we'd all missed, and he'd arrive at a diagnosis quickly.

I can still hear his sharp remarks in my memory. "Listen!" he'd snap at us, looking sheepish in our scrubs wrinkled from not having slept the night before. "You never listen!" He was always right, because if we had listened to the patient a little bit more, he or she would have handed us the diagnosis on a silver platter.

The art of diagnosis applied especially to the patient with acute abdominal pain. Timing was everything, and Dr. Silen was adamant about this simple dogma. Many people with the most severe belly pain will try to tough it out during the day, only to arrive in an emergency room late in the day or evening. Dr. Silen would remind us that just because someone shows up at night gives us no reason to press pause and see how things are the next morning. If symptoms had been going on for six hours or more prior to arrival, decisions have to be made and acted upon. "Sleeping on it," as he would warn, could be life threatening.

Gone are the days of asking the right questions, doing just the right amount of listening, and timing things perfectly. Appendicitis is no longer diagnosed by a careful history and a meticulous physical examination, complete with attention to associated subtle signs (e.g., the psoas sign, whereby touching the abdomen causes a muscle to flinch so that the leg elevates). Another old-fashioned way to diagnose appendicitis is to look for rebound tenderness or percussion tenderness, where it's not the pushing down on the belly that hurts, but it's the release of the palpation or transmission of touch that triggers the pain—a sign of peritoneal inflammation, a sign pointing in the direction (although not quite giving a clear diagnosis) of appendicitis. The peritoneum is the layer of tissue that encases the abdominal organs, including the appendix. Normally, these organs are floating in a bit of fluid, all encased by the peritoneum. If an organ either becomes inflamed or ruptures, the peritoneum becomes irritated. This is exquisitely painful, leading to abdominal muscular rigidity, and irritation if the peritoneum is further poked or prodded in any way. Peritoneal irritation was one of the most classic signs of abdominal organ damage. Now we don't need to assess peritoneal signs because we have CT scans! When a patient comes to an emergency room with acute abdominal pain, a CT is ordered immediately, and the pricey test is often completed before the surgical team gets the name of the patient, let alone a history and physical.

Before routine use of CT scans for belly pain, the concern was that we would wait too long to take a person to the operating room for appendicitis, especially if the H&P was inconclusive. Such delay put the patient at risk for a ruptured appendix, and its associated complications of bowel injury and widespread abdominal infections or sepsis. The other side of the coin was that patients were being operated on unnecessarily: Healthy appendixes were removed when the cause of the symptoms was gastroenteritis or a ruptured ovarian follicular cyst, a variant of normal ovulation in a menstruating female. Even something called mittelschmerz, or midcycle pain after ovulation, can lead to peritoneal irritation, looking just like acute appendicitis on a physical examination of a woman of childbearing years.

The goal of CT scans was to rid us of these misdiagnoses, or *negative appendectomies* (removing a normal appendix). In theory, if an abdominal CT scan shows an abnormal appendix, bowel obstruction, or inflamed peritoneum, out it must come. And if the scan doesn't, in it stays. But what if the CT scan is done too early, prior to the appendix's becoming inflamed? Get a second or even a third CT? A ten-year review of over thirteen hundred adult appendectomy patients, published in 2010, showed a fivefold increase in use of preoperative CT scanning for diagnosis, from 18 percent in 1998 to 94 percent in 2007.[9] Despite this marked increase, little to no decrease occurred in negative appendectomies in these patients. While CT scanning holds minimal risk, was this extra test worth it? No clinical recommendation can be based on one (or even two) studies, no matter how large. However, a second study, analyzing the negative appendectomy rate, or NAR, found that over eighteen years coincident with substantial increased use of CT scanning for diagnosis (from 1 percent to 97 percent), the NAR dropped from 23 percent to under 2 percent. The annual number of appendectomies was cut in half.[10] Seems clear, right? CT scans are the way to go. Look how many unnecessary surgeries were avoided! But, at the time of these studies, other studies were evaluating whether surgery was needed even when a diagnosis of appendicitis was confirmed. In many cases, even greater than half the cases, appendicitis will resolve with antibiotics, obviating need for surgery.[11]

I don't mean to imply that we don't know how to best evaluate and manage abdominal pain. Nor am I insinuating that we have gotten nowhere in fine-tuning evaluation and treatment of such a common problem. Quite the contrary. Much to the dismay of Dr. Silen, who is still alive at this writing and very likely has held his ground on minimizing unnecessary testing, we've reached new heights. But much to Dr. Silen's delight, there is no one algorithm to follow (Dr. Silen notoriously despises algorithms). CT scans are useful with acute abdominal pain in the emergency room. But they are by far not the end to the story for evaluation or treatment. Many excellent physicians continue to rely on the H&P, with superb patient outcomes. Other testing continues to be a part of the evaluation—every female of childbearing age (in our hospital it's arbitrarily set at ages ten to fifty-three) with abdominal pain gets a pregnancy test, by either blood or urine. A tubal pregnancy, also known as an ectopic pregnancy, can look like appendicitis. Elevated white blood cell count, although somewhat nonspecific, is an indicator of an acute infection. Some hospitals have ultrasound machines in their emergency rooms, which are in some ways safer than CT scans, as no radiation is given. Ultrasounds are not as accurate as CT scans, but they may reveal an enlarged appendix or certainly an abscess or fluid collection in the abdomen.

TO SCREEN OR NOT TO SCREEN

The second-leading cause of cancer death in the United States, after lung cancer, is colorectal cancer. Approximately 130,000 people are diagnosed annually. Nearly 50,000 will die of it.[12] As is the case with many cancers, colon cancer is treatable, and curable, with early diagnosis and intervention. Screening is key, as most early colon cancers present with no symptoms. So it seems obvious that the take-home should be to screen early, and screen often. For the most part, at least for screening early, this recommendation holds. The caveat is that as people age and/or develop other medical issues, such as heart disease, high blood pressure, diabetes, or respiratory issues, the risks of colonoscopy may make it not worthwhile.

One study reviewed adverse events over three years (2006–9) in three states (California, Florida, and New York) following colonoscopies.[13] The study looked at records of over 4 million adults, ages nineteen to eighty-five, and found that as patients age and/or have other medical problems, risks of intestinal injury or bleeding increase enough that it may be more prudent not to do the test, unless there is a strong suspicion of cancer. As of 2016, the U.S. Preventive Services Task Force updated recommendations that all asymptomatic people aged fifty should have a colonoscopy. If the results are normal, a second one should be done at age sixty and a third at age seventy. After age seventy-five, whether to perform a screening colonoscopy should be decided individually—for instance, if there has been a history of colonic polyps or if there's a strong family history of colon cancer.

Nothing in medicine, including published guidelines for colonoscopy or any other screening procedure, is straightforward. And it certainly is never the final word. First of all, no randomized clinical trials have been conducted to determine whether colonoscopy screening at ten-year intervals reduces the mortality from colon cancer. That's right, it's a recommendation without the purest research evidence that it reduces mortality. This snafu comes into play for many procedures or recommendations that have such strong positive results. It would be difficult to create two large groups of fifty- to seventy-year-olds and randomly have half of these patients *not* have colonoscopies (and see who dies from colon cancer) while the other half do have colonoscopies. However, some trials are in progress that compare screening colonoscopy to other colon-cancer-screening techniques, which I'll describe below. The final results of these trials may not be known for several decades, but even preliminary results show that colonoscopy is preferable to other screening methods to reduce mortality.[14]

Being a doctor means that I'm friends with other doctors. Not only is it said that doctors make the worst patients, but we are also notorious for having a higher risk of complications. At least that's how we see it. "Doctor's kid," we'll mutter in unison as an OR team as we wheel in a three-year-old for surgery. These two words have so much meaning: Be extra careful—the parent will notice anything. Give VIP treatment. Bump the patient to the front of the line on a long surgery day. Bring in the A-team. But what we're really thinking, first and foremost, is *increased risk for complications*. While this is wholly based on superstition, ask any doctor what it's like to treat a doctor, or the family member of a doctor, and we all cringe. Not because doctors are more demanding or not compliant. In fact, the opposite is usually the case. If anything, we tend to be more understanding of scheduling delays, insurance glitches, and even complications. But a common notion (albeit false) is that if something bad, no matter how rare, is going to happen in a medical setting, it's going to happen to someone in the medical field. So of course when a doctor friend has a colonoscopy, she will be the one young, healthy person to have a rare complication. Which she did. Followed by complications of treating the complication. Twice. Which leads me to wonder, are there other methods for screening for colon cancer, perhaps just as accurate but with fewer risks? Of course.

One of the least invasive (unless you're the patient) tests that we learn in medical school is what used to be known as the stool guaiac test, which we performed on every adult undergoing either a routine physical examination or admission to the hospital for any reason. A good old-fashioned rectal exam. This is now known as the FIT (fecal immunochemical test) for occult blood and the gFOBT (guaiac fecal occult blood test). Even more specific is the FIT-DNA test (multi-targeted stool DNA test). As you'd imagine, the mechanics for obtaining material for these tests is with a gloved finger in the rectum. The material (even if it's just clear fluid, without feces) is placed on a card with an indicator for + or -, indicating presence or absence of hidden (occult) blood—blood that you cannot otherwise see. Every medical student is familiar with the stool guaiac test as a critical part of any physical exam. An intern's job is to perform a rectal exam and stool guaiac on every medical, surgical, and trauma admission. While the term *guaiac* seems like a medical one, it's actually derived from the brown resin, used for flavoring (yum!), from the guaiacum tree.

The presence of occult blood leads to an exhaustive differential diagnosis, including internal hemorrhoids, fissures, history of constipation, history of anal intercourse, and benign and malignant colon lesions. As the list is long, a positive test merits further evaluation, either by flexible sigmoidoscopy (mini-colonoscopy) or colonoscopy. CT colonography, or virtual colonoscopy, is another screening tool, avoiding the risks of colonoscopy, but adding risks of radiation from CT scanning, finding of extracolonic abnormalities (this may actually be a good thing, although it can lead to further unnecessary testing), and lack of guidelines regarding how often this test should be performed. Additionally, if an abnormality is found during this procedure, you will have to undergo a colonoscopy to biopsy the mass or polyp. This could have been done at the time if a colonoscopy had been performed.

We are always concerned about finding things we're not looking for—a CT scan of the sinuses for sinus infections, for instance, may show a benign brain abnormality, leading to further testing and even a brain biopsy for an inconsequential lesion. When we're stuck with the information, we have to act on it. Occasionally, this is for the good. One of the most well-known instances of this occurred with Supreme Court Justice Ruth Bader Ginsburg. She had been treated for colon cancer, and during a routine follow-up abdominal CT scan, a small pancreatic mass was found. This was cancerous and was removed successfully in 2009. The vast majority of pancreatic cancers do not rear their ugly heads until symptoms set in at an advanced stage of the disease, which is one of the reasons that this cancer has such poor outcomes. By the time most pancreatic cancers are discovered, the disease is too established to cure. But because Justice Ginsburg's cancer was found incidentally, it was removed before it progressed to invasive disease. Had Ginsburg not had that screening CT scan (looking for colon cancer—not pancreatic cancer), in all likelihood her pancreatic cancer would have progressed, with a much different outcome. Let's hear it for screening scans!

Shouldn't we all sign up for full-body CT screening scans, or even PET-CT scans, which give an even more accurate view of cancerous growths? The answer is no. CT scans for no other reason but screening alone have been found to do more harm than good. Even the American College of Radiology recommends against these, for the very reasons you'd think: unnecessary radiation, and finding of those pesky incidentalomas that lead to unnecessary procedures. Ginsburg's case was a rare, lucky finding.

Even less radioactive tests may be a waste. The American College of Preventive Medicine does not recommend electrocardiograms, stress tests, or blood tests for inflammation in patients with no cardiac symptoms. On the other hand, blood pressure screening, cholesterol screening, and Pap smears in women under age sixty-five are a big yes for their benefits. The PSA, or prostate-specific antigen test, is considered the gold-standard marker for early detection of prostate cancer. In 1992, the American Urological Association and the American Cancer Society both recommended annual PSA screening for all men over age fifty. With this recommendation came a sharp increase in the reported incidence of early-stage prostate cancer. No surprise. You test everyone, you're going to find more asymptomatic disease, just as one would performing colonoscopy. The difference is that the PSA test is a blood test, with virtually no procedural risks. The risks come later, associated with biopsies, surgeries, and radiation therapy. Colonoscopy with or without biopsy carries procedural risks, with benefits of early treatment far outweighing these risks. This is not the case for PSA screening, so much so that in 2008, the U.S. Preventive Services Task Force recommended against PSA screening for men over age seventy-five, and in 2012 against it for all men, even those between fifty and seventy-five. With these changes has come a substantial decrease in reported incidence of prostate cancer, from over 210,000 diagnoses in 2011 to 180,000 diagnoses in 2012.

While it may seem that thirty thousand men are walking around with undiagnosed prostate cancer, prostate-cancer-related deaths have not risen.[15] As one of my pathology professors in medical school once said, more men die with prostate cancer than die of prostate cancer. Just as mammograms may or may not be indicated on an individual, case-by-case situation, PSA testing may or may not be used routinely, but should be discussed by physicians with their fifty-five- to sixty-nine-year-old male patients. Each man can then incorporate the information regarding the benefits and risks of this screening exam for his individual situation.

For those with good health insurance and finances, myriad screening tests can be done to take a better look at your innards. I'm not talking about getting an MRI of the brain to see whether your migraine headaches are just migraines, or a chest X-ray to evaluate for pneumonia. I'm talking about pure screening tests. Just because. Because you can. Because they are there. The marketing for these can be incredibly enticing: "To prevent premature death." "To detect hidden diseases." Words such as *noninvasive, low radiation,* or *mini-radiation* will get the skeptical, deep-pocketed consumer in the door. These tests do offer the highest grade in technological advances. A CT coronary angiogram can measure how much calcium is present in the coronary arteries and give the patient a calcium score, which can indicate a silent plaque buildup in the arteries, which puts the patient at risk for heart attacks or strokes. And premature death.

You can also go for broke and get a full-body (chest, abdomen, and pelvis) CT scan, to find those hidden time bombs waiting to explode. Causing premature death. An MRA is a combination MRI and angiogram (a look at blood vessels) to check for aneurysms and vessel abnormalities in the neck and brain, which could be a setup for a stroke or ruptured aneurysm. This would, you guessed it, lead to premature death. A "low-dose" lung CT looks for asymptomatic lung cancers. Preventing premature death.

All of these exams are high quality, but are problematic when they are ordered only for screening. Most people who hear devastating stories about sudden death in previously healthy people would think that if only they had been screened for (fill in the blank), they would still be with us today. Unfortunately, from both a public health, and even an individual health, standpoint, this is just not practical. Yes, good preventive health initiatives include screening tests, many of which, such as mammography, remain controversial. But the notion that a screening test will prevent all of these premature deaths is unfounded. It would be nice, but it's not the case. This takes us back to dear Dr. Silen. Every test needs an indication, and tests should only be done if you plan on acting on the results. Dr. Silen would go ballistic if we ordered routine labs or post-op blood tests just because it was our robotic routine for hospitalized patients. We were forbidden from treating a fever (that's right, not even with Tylenol) until we found the source. Every test has to be done for a reason, and that reason can't be because it's there.

What still strikes me is how testing is often taken more lightly than the potential heavy implications of results. Many don't consider the what-ifs when blood is drawn, butts are swabbed, urine is collected, or scans are completed. Families with heritable diseases, such as Huntington's disease, breast and ovarian cancers, or sickle-cell anemia or other blood disorders, know all too well the implications of testing. This is why many choose not to test—they know that the results may change their lives, long before any symptoms rear their ugly heads. My colleague who has three children, each with a 50 percent chance of having Huntington's disease, will not test them. Knowing such information at this point will only change their knowledge; it will not stave off the disease.

I have another friend whose mother had breast cancer. Although her mother was treated and survived, this friend, now in her fifties, has never had a mammogram. Testing is scarier when one fears the results. A medical school classmate is a radiologist who specializes in breast imaging. She, too, also in her fifties, has never had a mammogram. Although she has no family history of breast cancer, and no signs of a breast mass, she fears the results of a test due to her daily views of abnormal results.

In medical school, students famously suffer from something we like to call "second-year disease." While the first year of medical school focuses on how the body works, the second year is spent learning about how diseases wreck us. We don't learn about treatments until we get to the clinical wards in the third and fourth years, so in second year, we are all somehow dying—of stuff mundane and obscure. My suntanned roommate was convinced she had a rare liver disorder, causing the skin to turn a copper color, called hemochromatosis—not that she was just glowing amber from her suntan. Upset stomach? It must be gastric cancer. Irregular periods? Can't be stress related, but must be a massive ovarian teratoma. Despite that our fears were misguided, most of us still underwent routine testing. Even though we were students, we were also young adults on our own, and it was time to be responsible.

I'll never forget the week or two after a routine Pap smear I had during my second year of medical school. I wasn't having any symptoms—just a routine test, right? But when one week, then a second, and then a third went by, I began to panic as to why the doctor hadn't called with the results. Something must be wrong. Maybe he's getting up his nerve to present me with the bad news. I must have cervical cancer at best, ovarian cancer at worst. Maybe it is inoperable. I made plans on how I would break the news to my family. But then I thought—maybe if I don't find out the results, my cancer will magically disappear. Ignorance is bliss, isn't it? At the fourth week, still with no phone call, I pulled myself together to call the doctor.

"What's the problem?" he asked curtly when his assistant put me through.

"I was wondering what my Pap smear showed."

"Oh, that? Normal. I only call if there's a problem."

"Okay. Thanks."

When getting a test done, make sure you think it through—normal results are usually just a quick call, or no call at all. But always ask about the path where abnormal results will take you. Those of us in medicine know all too well about these paths, as do those whose families have had illnesses being screened. But most healthy folks can't possibly know the implications of every blood test, X-ray, or exam. Which is why discussion about the test is in order before any test is done.

HYPE ALERT

* Most everything in medicine comes with potential upsides and downsides. Not everyone who takes a certain medication, for example, will experience a negative side effect, and not everyone who takes the medication will benefit from it.

* Often, nothing beats a classic history and physical (asking questions, examining the body in person)—even when high-tech diagnostic testing is available.

* CT scans for no other reason but screening have been found to do more harm than good. Electrocardiograms, stress tests, or blood tests for inflammation in patients with no cardiac symptoms are not recommended. But blood pressure screening, cholesterol screening, and Pap smears in women under sixty-five are a big yes for their benefits.

* More men die with prostate cancer than die of prostate cancer. Just as mammograms may or may not be indicated in individual, case-by-case situations, PSA testing may or may not be used routinely, but should be discussed by physicians with their fifty-five- to sixty-nine-year-old male patients.

12

WHEN 50 IS THE NEW 40: DRINKING FROM THE FOUNTAIN OF YOUTH

How to Age Gracefully, Without Really Trying

WHAT'S THE BEST ANTIAGING TRICK OF THE DAY THAT'S NOT HYPE?

WHY IS SPF 100 WORSE THAN SPF 30?

IS TESTOSTERONE SUPPLEMENTATION A RISKY ENDEAVOR
OR WORTH IT FOR QUALITY OF LIFE?

IS POSTMENOPAUSAL HORMONE REPLACEMENT SAFE
UNDER CERTAIN CIRCUMSTANCES? WHAT ABOUT
BIOIDENTICAL HORMONES?

WILL WE EVER BE ABLE TO LIVE TO 150?

Since time immemorial, humans have searched for ways to look and feel younger, if only temporarily and with risks and costs. The fountain of youth—a natural spring that restores youth to anyone who drinks or bathes in its waters—was first written about in the fifth century BCE. In medieval times, women bleached their skin using lead paste. During the Renaissance era, Catherine de' Medici used pigeon poop on her face to get that dewy, young complexion. Mary, Queen of Scots, kept herself looking youthful and fresh by bathing in wine. Today some people will go under a knife or inject themselves with botulinum toxin to turn back the clock. The number of medical spas, hybrids of medical clinics and traditional day spas, in the United States is up more than fourfold since 2007.

Perhaps no other industry has duped us more and taken more of our money than the antiaging business. Claims of miraculous remedies and therapies to overcome aging run rampant, despite that many of these products have no lasting effects whatsoever. Firming cream. Line eraser. Wrinkle cure. Body-slimming wraps. Chemical peels. Injectables. Implants. Hype. Hype. Hype. Some coveted moisturizing creams, which tend to hide their "secret" ingredients, retail for hundreds of dollars per two ounces.

To be fair, some of this stuff works, but the operative word (and I mean *operative* both literally and figuratively) is *temporarily,* and sometimes in an illusory fashion. Chemical peels can strip the layer (or layers, depending on how far you go) of dry, dead, wrinkly skin to reveal fresh, glowing, youthful raw flesh. As a younger surgeon, I often performed dermabrasions—using an instrument that is a cross between a dental drill and an electric toenail shaver to scrap off layers of scars, wrinkles, and age spots. The result would be a lustrous, raw, burned face, similar to a veritable traumatic-burn victim's. But in a few weeks, voilà! Beautiful skin. Temporary beautiful skin, that is. Until the real stuff grew back. The pockmark scars from adolescent acne would be obliterated indefinitely, but wrinkles would return in due time. No skin cream can reduce wrinkles; creams just minimize the *appearance* of wrinkles. A big distinction.

The antiaging industry is widely studied by both medical science (what truly works?) and popular psychology (what if you *think* it works?). Let's take a tour of the biggest offenders in the antiaging schemes and hoaxes category. But I'll also offer some scientifically backed advice on how to look and feel younger.

With new data showing that many babies born in this millennium will live well into their hundreds, we are more and more eager to turn back the clock, take off ten years, and feel like a kid again. If fifty is the new forty, and forty is the new thirty, are we really aging differently from generations past? My great-grandmother lived to be 104. She was born in the late 1800s, and had a tough life. She moved from

Eastern Europe to England, then to the States, all before age twenty-five. She sewed for a living and was of the first members of the ILGWU (International Ladies' Garment Workers' Union, one of the first U.S. labor unions). She wasn't educated, and she worked in factories all of her adult life. She didn't exercise, except when she mashed raw fish in the bathtub with a baseball bat. She ate a lot of fat and was pretty hefty throughout her life. She had three children who survived to adulthood, and several who didn't. She had many miscarriages. Her oldest son, my grandfather, died at age sixty-six—heart disease, obesity, high blood pressure, and a ruptured thoracic aneurysm led him to a precipitous death. She lived twenty more years, during which time her younger son died. Her daughter outlived her, but did not nearly reach triple digits and succumbed to a rare cancer.

How did this woman, born more than a century ago, when few of the medical advances of today existed, outlive most of her children? One might say good genes and good clean living, but is either really the case? A dear friend's husband—thin, healthy, exercise maven, didn't smoke, ate right, didn't drink—died of pancreatic cancer at age thirty-seven. Bad genes but good clean living? We may never have the answer. Even as a doctor, I can't predict who will live long and who will likely have a short life. I've known people who circle their drains for years in terrible health while others are snuffed out much earlier than expected. Probably just as many people who engage in the worst lifestyle habits live into their nineties as do people who try to live impeccably and who die prematurely (though even that word *prematurely* is suspect!).

Many of us are now realizing that creams, lotions, potions, lasers, and even plastic surgery do not keep us young on the inside. *Lifestyle* tends to be a new buzzword. Lower your stress. Turn off your smartphone. Get some sleep. Exercise, but not too much. Eat some bran (but not too much bran!). Go back to sleep. Wake up. Go to work, but don't work too hard. The perfect balance is always a goal, but nobody can claim they have found it. Why would my lard-eating great-grandmother live to be 104, and my friend's svelte soul mate die at 37?

It's one thing to look ten years younger and quite another to actually *be* ten years younger physiologically. We don't know how or when we will die, and perhaps that's a good thing. We'll keep doing our best to look and feel as young as can be, and there's nothing wrong with that. We might not be able to push our longevity that much further (we may never live to two hundred), but we can delay the onset of disease. So we can enjoy more years of a healthy life before succumbing to an illness or disorder (that hopefully takes us out quickly). As you read this, drugs such as metformin, the most common drug used by type 2 diabetics and which has been around for sixty years, are being tested on healthy humans to see if they can extend people's healthy years. And this isn't hype. It's promising.[1]

CALIFORNIA DREAMING

When I first moved to Los Angeles, after years in the Northeast, a friend who's a California native told me, "You'll feel ten pounds heavier and ten years older as soon as your bags are unpacked." In just a short time I figured out what he meant. Southern California is a land of extremes: products, injectables, surgeries, "health food" stores, exercise and weight loss innovations, and desire to drink from the fountain of youth. Everyone looks thinner and younger than people do in other cities across the country. The age discrepancies in couples seemed more dramatic here than anywhere else I'd been, and I often wondered if a wife's age in years was more or less than her husband's PSA level. Many of these beauties looked darn right strange, but many looked amazing. What was their secret? Was it just that they were twenty-five, or were they fifty, looking twenty-five? And if they looked so good, what was needed to maintain such firm skin? Lotions, potions, or something more?

During internship and residency, we had the expression "They [meaning, our mentors and teachers] can always hurt you more, but they can't stop the clock." In other words, time would keep ticking, and we would eventually be finished—with a hard night on call, a tough rotation, and the year, no matter how painful our work days and nights were. The latter part of this phrase, like it or not, holds for youth: you can't stop the clock. When it comes to skin, this is absolutely the case. Genetic as well as external factors may slightly alter the speed of the clock, but it will tick on nonetheless. For instance, darker-pigmented skin tends to be heartier; poor nutrition and chronic systemic diseases may have a negative impact on skin integrity; and my old frenemy, my most self-harming indulgence, sunshine, does the most damage at any age, speeding up that clock with abandon. Moisturize, plump, hide, conceal, but your skin will continue to age as it was meant to.

The skin, the largest organ in the body, is made up of three layers: the epidermis, the dermis, and the subcutaneous tissue. The epidermis is the outermost component, consisting of cells, pigment, and proteins. The dermis is just underneath and contains blood vessels, nerves, hair follicles, and oil glands. The deepest, or subcutaneous, layer contains fat, some hair follicles, sweat glands, and blood vessels. The tightness of younger skin is due to the presence of collagen in all of these layers, and elastin for elasticity. With age, the epidermal layer thins, the pigment cells dissipate, but what's most notable, and what so many are trying to halt or reverse, is the gradual loss of connective tissue, primarily collagen, the main structural protein in the skin, which provides its integrity.

The simplest, and most common, way to prevent loss of the skin's luster is to avoid the sun and to not smoke. These two practices are the foolproof, least expensive methods to maintain youthful skin. When smokers undergo cosmetic surgery (and many plastic surgeons will not operate on smokers), the healing time doubles, and the risks of skin breakdown and wound infection rise substantially. The number one cause of skin damage, including the hastening of the aging of all skin layers, is exposure to UV light. Wearing sunscreen regularly, even one that is incorporated in daily moisturizers, is great, even on cloudy days. Staying out of the sun can't hurt, either.

My medical school classmate Dr. Andrew Sussman is the executive vice president and associate chief medical officer at CVS Health Corporation. So if you are frustrated that you can no longer buy cigarettes or sunscreen with less than SPF 15 at your local CVS, you can thank a doctor. While he acknowledges that not selling cigarettes will not necessarily lead many people to stop smoking, preliminary data does show some promising numbers—people who are at the pharmacy shopping for medications to treat chronic illnesses that would be exacerbated by cigarette smoking have been thinking twice and cutting down on tobacco use.[2] A small secondary benefit would be healthier skin. At a recent medical school reunion, poor Andy was lambasted by one of my fellow sun-worshipper classmates for her not being able to buy SPF 4 or 8 at her local CVS anymore. That darned health-conscious chief medical officer was blocking her sunblock choice.

Many also swear by skin creams, or moisturizers. Most of these are over-the-counter, and prices can range from under $10 to thousands. Is this price differential worth it? Will designer moisturizers give you designer skin? Claims on some of these labels sound as good as a J. Peterman catalog's elaborate description of a drab winter coat: "This award-winning technology . . . effectively transports potent ingredients to where your skin needs it most." "Scientifically formulated to help minimize the appearance of wrinkles." "Skin looks virtually ageless." That's right, you can have ageless skin for a mere $75 to $150 per transported ounce. Or, you can have "visibly smoother and younger looking skin" for one-tenth the price. Which is better, if any of these are useful at all? Can they reset the clock?

As science advances, we have more and more methods of halting loss of collagen, or stimulating collagen growth or strengthening it, enabling patients to avoid the knife or even the needle. Some of these potions, containing such ingredients as retinols, peptides, and growth factors, do, indeed, go more than skin deep, entering the dermis, where the real action of tensile strengthening and repair takes place, stimulating collagen production. Antioxidants such as vitamin C, B_3, and E can penetrate the skin and reduce collagen breakdown.[3] But for the most part, especially in over-the-counter creams that no regulatory agencies monitor for claims and labels, it's all a lot of smoke and mirrors. The FDA is responsible for monitoring the safety of these substances, but not their efficacy.[4]

Some of the popular ingredients you'll see in the skin creams are antioxidants, those magic substances that somehow bind up evil free radicals all over the body. Such ingredients include retinol, vitamin C, niacinamide, tea extract, and grape-seed extract. Hydroxy acids act to exfoliate, removing dead skin cells. Peptides and coenzyme Q aid in wound healing and protection from sun damage. Some of the products with these ingredients can make a difference; your skin will look healthier (translate: younger). But there are several caveats, and here are just a few: Using more than one ingredient does not necessarily double or triple the benefit. Some ingredients are totally inert once you smooth them over your face because air and light deactivate them (or they were poorly packaged). Many of these require daily or twice daily use indefinitely, with skin's return to its leatherette quality soon after stopping regular use. Cost does not equal benefit. Some products may lead to nasty skin reactions. And to be politically incorrect, all skin is not equal—what works for your friend may not work for you.

Georgia O'Keeffe, renowned artist, who lived to the ripe old age of ninety-nine, had beautiful, deeply set wrinkles, mimicking the cracked earth of the desert in which she resided for so many years. In 2014, then sixty-eight-year-old Diane Keaton was happy to use and promote L'Oréal skin cream, enjoyed good lighting and touch-ups on photos, but was completely uninterested in permanently altering her unstoppably developing wrinkles. Look at photos of Helen Mirren, Maya Angelou, or Golda Meir. These women are institutions who aged gracefully, wrinkles and all. As with so many measures of beauty, there is the gender double standard—stars such as Robert Redford, Sean Connery, and Morgan Freeman can seem even more appealing (more distinguished) with a little skin laxity around the edges. So with all of the goodness, or sham, that skin products provide, they do only go skin-deep, providing no clock-stopping mechanism under the very thin surface.

Setting aside the magic of youth-restoring or youth-preserving creams, prevention is key for skin health. While one cannot stop aging, one can minimize the collateral damage. Sun exposure not only ruins skin, it is also responsible for at least ten thousand deaths annually in the United States alone. One person dies every hour from melanoma, the deadliest form of skin cancer. Daily use of sunscreen of SPF 30 or greater can reduce this risk by up to 50 percent.[5] So if we can cut the risk by 50 percent using SPF 30, why not use the highest SPF possible? SPF 100, or even higher? The answer is, it doesn't help. SPF stands for "sun protection factor," but the number represents time, not strength. For instance, if you are wearing SPF 30, you can be in the sun thirty times longer before incurring the same sun damage as if you were wearing no sunscreen. Again, so why wouldn't you go for the SPF 75 or SPF 100? The sun emits two different types of rays that damage the skin: ultraviolet A (UVA) and ultraviolet B (UVB). The latter is the most responsible for skin cancer and sunburns (as it is said in dermatology circles, A is for aging and B is for burn). Most sunscreens provide SPF based on UVB protection. However, UVA rays also contribute to skin cancer, but are not always blocked in sunscreens. Those wearing SPF above 50 will burn less, but be exposed more than those with lower SPF, giving a false sense of protection and actually increasing the risk of cancer. Most, but not all, sunscreens do contain both UVA and UVB protection, which is more critical than the SPF number.[6]

Despite our best efforts at sun protection, skin changes do occur naturally with age. One step beyond skin creams (topicals) is a chemical peel. In this treatment substances such as glycolic acid and TCA (trichloroacetic acid) are applied in liquid form to the face. The varying strengths and durations for use depend on the desired depth of penetration. A peel does just as you'd guess: it peels one or more layers of the skin off, creating a superficial or deep sunburn effect, without the ultraviolet radiation, enabling new skin to grow in its place. Peels have been used for years—once only in a doctor's office, then at the aesthetician's, and now in one's home. Risks are primarily deeper burns, allergic reactions, and permanent scars, and benefits vary from no change to brighter skin. Results last from weeks to months, but are not permanent. Skin creams and peels, no matter how advanced the technology, can only go so deep.

When it comes to aging, gracefully or otherwise, the skin reveals the first visible signs of the passage of time. Taking care of one's skin, whether by maintenance or therapy, is vital to one's outer appearance, which carries over to one's inner self-confidence. However, there is a difference between looking better and looking younger. Skin rejuvenation can most definitely make skin look better. Smoothing out blemishes, filling or erasing scars, or removing darkly pigmented areas all give a better look to the skin. But it's important to keep your goals in perspective—are you trying to look healthier or younger? Maybe both!

THE PRETTY IN PARALYSIS

When I was young, we were told never to eat soup from a can whose metal bulged out. When I was older, I learned that infants under age one were not to eat honey. Both of these tenets to put the fear of death into us were for the same reason—botulism. So when my then ten-month-old son grabbed a handful of Honey Nut Cheerios, my husband and I, being the difficult-to-frazzle doctors that we are, called our pediatrician in a panic. All because of botulinum toxin. No, not the wrinkle remover known as Botox. The real toxin. Botulinum toxin is a neurotoxic protein produced by the bacteria *Clostridium botulinum*. It is one of the most potent neurotoxins on the planet, blocking the release of acetylcholine at the neuromuscular junction. The blockage of this neurotransmitter leads to muscle paralysis. Infectious botulism is usually food-borne or from an infected wound, leading to full-body muscle paralysis, respiratory failure, and possibly death. The treatment is an antitoxin and mechanical ventilation. While infection is now rare, infants under twelve months are susceptible if they eat honey, which may contain minute amounts that can be easily fought off by older toddlers.[7] Little did my husband and I know that Honey Nut Cheerios doesn't contain much honey to speak of. And even if it did, the factory processing was such that any organism, even botulinum, wouldn't survive. And the soup-can fright? Same reason. The bacteria, known to grow in expired canned products, release a gas that can bulge out a metal can. A bulging can may be a sign of active botulism, so stay away. And the obvious next step is to inject that stuff! Everywhere!

Botulinum toxin, or Botox, is now used in virtually every medical specialty.[8] It was popularized in the nineties as a wrinkle blocker, or miracle poison, and Botox parties with countless Dr. Feelgoods followed. Botox is now used to treat migraine headaches, chronic pain, muscle spasm, vocal-cord dysfunction, hypersalivation, facial twitches, and many other ailments. The list continues to grow. But what it's most known for is wrinkles. It was first used in the brow, in the glabella, right between the eyebrows, to paralyze that muscle and get rid of frown lines. Then came the forehead wrinkles. Then the crow's-feet around the eyes. Then the deep lines surrounding the nose and cheek. Then the smile lines around the mouth. Then the neck. And so forth. Once the rage began, people thought that Botox wasn't doing enough, that once a wrinkle formed it was too late for the injection. So began prophylactic Botox—to prevent wrinkle development. Paralyze that muscle before the wrinkles can grow. Kids (yes, kids) as young as in their late teens jumped on the Botox bandwagon. Botox—and its siblings (Dysport, Xeomin)—continues to be the injection of choice. It's fast, relatively painless, with virtually no recovery time. There is a bit of a lag before one sees results—a few days to a week, but most feel it's worth the wait. The results last from three to six months.

Sometimes youth seekers are not after wrinkle flattening but skin plumping. Even the most advanced topicals (creams, lotions, gels, and peels) will not plump up the skin to one's desire. Ah, but there are injections for that, too! Collagen is the mainstay, with many iterations of this magic substance over the years. Some to plump up the lips, fill in the folds, and smooth out the wrinkles. Some to plump up the cheekbones. All of these are temporary, also lasting three to six months.

For those seeking a more permanent tight face, there's always surgery. Traditional face-lifts, where an incision is made around the ear, or along the hairline, to hide scars, pull back skin, muscle, and the layers around these to tighten them. Only masterful plastic surgeons can accomplish a natural look with such a drastic operation, so as not to give the driving-a-Porsche-with-the-top-down windswept look. But even face-lifts have become less invasive, using either tiny telescopes or needles and sutures alone, with no incisions at all. This skin tightening, trimming, and cutting is temporary, too. It may cut years off one's appearance, but all face-lifts eventually fall. As a wise plastic surgeon once told me, plastic surgery is the one specialty where you can make a healthy person sick. Undergoing surgery at an advanced age carries risks, more so than in younger people. Risks of anesthesia, especially general anesthesia, increase with increasing age. Long-term effects to the brain may occur with general anesthesia in those over sixty-five, especially if it is prolonged (over three hours, which many plastic surgeries are). So undergoing such procedures to create a visage of youth may actually age you. After all, aging is irreversible. Or is it?

SUPERMAN AND WONDER WOMAN

While we know that all the nips and tucks from head to toe are literally skin-deep, what about digging a little deeper? Are there youth hormones that go dormant with age, but can be replaced with a pill or a gel? Well, yes. And no. As men age, the level of the most commonly known sex hormone, testosterone, begins to wane. With this comes reduction in sexual desire, physical stamina, energy level, and muscle mass. Testosterone levels are easily assessed by blood testing, and testosterone replacement has been a common fix for men whose levels are genuinely too low. Some data have shown that testosterone therapy increases the risks of cardiovascular disease and prostate cancer, but more recent studies show no significant negative impacts. In fact, it actually helps. Men with low testosterone levels who receive supplementary testosterone report increased energy, sexual desire and function, and physical stamina with exercise. Risks of therapy were not any more significant than those incurred simply by being male and over age sixty-five. The likelihood of cardiac events or prostate cancer was not greater with or without testosterone therapy.[9]

That said, those who take off-label testosterone, which may not be regulated for safety, purely to increase energy and stamina, with no monitoring by a physician, run into big trouble. As with any product, obtained either by prescription, online, or at a health market, more does not necessarily mean better. This is certainly the case with testosterone supplementation, where too much can absolutely increase risks of cardiac complications, including heart attack and stroke. And, no, superdosing on testosterone will not make you Superman. It can in some cases turn what would be treatable cancers into widely metastatic diseases that are difficult if not impossible to rein in. This is especially the case when testosterone is combined with HGH, or human growth hormone, another so-called wonder drug or fountain of youth. While it is unclear whether taking HGH *causes* cancer, it has been found to make precancerous cells more likely to become cancerous, allowing these previously dormant cells to grow rapidly.[10]

What about Wonder Woman? Estrogen is the primary sex hormone for women, and it's often associated with youthfulness and sexuality. Estrogen replacement therapy is not new, but it has certainly been fraught with controversy and risk. As women enter menopause, anywhere from the late thirties/early forties to late forties/fifties, estrogen and progesterone levels fall. This not only affects sex drive, fertility, and menses, but also bone density, fat distribution, cardiovascular health, and cholesterol. Estrogen has often been thought of as a protective hormone for women—guarding against heart attacks, high cholesterol, and bone weakening/fractures. Cholesterol comes in several forms, and the good kind, high-density lipoprotein (HDL), is thought to offer cardiovascular protection. Women tend to have higher HDL levels than men, and this is due to endogenous estrogen—estrogen naturally produced in the body. When estrogen levels go down, so do HDL levels. Many studies show that estrogen replacement therapy does, indeed, raise HDL levels and lower LDL levels.[11]

But replacement is not that simple.[12] While some studies have shown that estrogen replacement, with or without progesterone, can lower the risks of heart problems, breast and colon cancers, and bone weakening and fractures, these benefits are short-lived, only lasting during therapy, and can often *raise* the risks instead of lowering them. For instance, while some data show that hormone therapy lowers the rate of breast cancer, the data also show that when breast cancer does occur, it's a more advanced form. This holds for colon cancer, as well.

Nowadays, doctors recommend that hormone replacement be taken for the shortest time possible (primarily for symptoms of menopause) and be stopped as soon as is tolerable.[13] What's come of all of these valid concerns is the emergence and sale of bioidentical hormone replacement therapy, or BHRT. The brilliant marketing idea is that these substances will perfectly mimic one's own hormones, identically matching them biologically. Sorry, not so. There is no credible evidence that these work nor are they regulated by any outside agency, including the FDA. Claims that these unregulated substances make you slimmer, prevent dementia, and are safer than traditional hormone replacement therapy are unfounded.[14] While certain antidepressants or even pain medications (gabapentin) may ease symptoms of early menopause, so-called phytoestrogens (plant estrogens in soy, herbs, and black cohosh) are not regulated, tested, or available only by prescription. They will either do nothing or some harm.

In 1991, the NCBI (National Center for Biotechnology Information, a branch of the National Institutes of Health) initiated a set of clinical trials of close to two hundred thousand postmenopausal women for over fifteen years.[15] In these primarily observational studies, one focus was whether hormone replacement therapy raised or lowered risks of particular illnesses. The long-term studies looked at association, not cause, meaning merely that one treatment was more likely to be associated with an outcome.

One group of researchers looked at women taking combined estrogen and progesterone replacement therapy compared to a placebo group. The combined-HRT group had increased risks of heart attacks, strokes, blood clots, breast cancer, and dementia in those over age sixty-five. The researchers found decreased risk of colorectal cancer and fractures, and no overall relationship to cognitive issues, in those under age sixty-five. In the group who took estrogen HRT alone, compared to those in the placebo group, the study found no increased risk of heart attack, but increased risk of stroke and blood clots. The association with breast cancer risk was not determined; there was no relationship to colorectal cancer incidence and no relationship to dementia or cognitive deficits. There was decreased risk of fractures.[16] In sum, there is no perfect recommendation on whether to take HRT. Each woman has inherent risks of cardiovascular disease, breast cancer, colorectal cancer, and fractures that need to be accounted for when making this decision.

When I was a kid, my friends lovingly called me Shrimpo. I was tiny. I looked five when I was eight. I was always picked last for kickball games, and I always sat in the front row for class pictures. I didn't mind it. Most called me cute, and I had no interest in being an amazon. I grew eventually, but I remained one of the smallest in high school, college, medical school, residency, and now. My daughter is small, but she cleverly calls herself fun-size. She got teased a bit in elementary school, but soon the kids encountered the wide range of body types at every age, whereby everyone is vulnerable, and the teasing stopped. Nowadays, parents are more and more concerned about height differences and are told early on the likely adult height of each of their children, even before a child is able to stand up on his or her own. Parents hold on to that number with pride. Unless the number is low.

Many reports have demonstrated that low adult height is associated with fewer job opportunities, lower income potential, fewer marital options, and overall dismal prospects in life. For girls, the cutoff is at about five feet—for boys, about five feet four. Any less than these, and life successes decline. We now have more accurate ways than the classic parental-height assessments to estimate how tall a child will be as an adult. Once any medical issue has been ruled out, an X-ray of the wrist bones will allow calculation of what's called a bone age. If the child's bone age is one to two years less than the child's chronological age, meaning that the body's physiological age is that much younger, then the lack of height may suggest the child is a late bloomer. But if the bone age matches the chronological age, and the child is well below the height of his or her peers, hormone levels may be checked by blood tests, and growth hormone therapy may be an option to add up to five or more inches to final adult height. Mind you, there is nothing wrong with being short. Typically, kids will start daily growth hormone injections at around age eight and continue them through adolescence.

My friend and colleague Dr. Cara Natterson is a pediatrician who consults with families whose child has either a new acute illness or injury. This might be a recent concussion or a bad pneumonia or a chronic illness fraught with associated challenges such as newly diagnosed type 1 diabetes or poorly controlled asthma. She helps the families understand the science behind the illness, helps them weigh the therapeutic options, provides answers to questions, and makes suggestions for further evaluations. Her hands-down most common consultation is about whether to begin growth hormone therapy for a child. And for good reason. Growth hormone is often indicated for children who are either short or have other quite rare syndromes, such as Prader-Willi syndrome. The benefits of daily injections of growth hormone are seen literally before your eyes in weeks to months, but the risks are real. If a child is not simply a late bloomer, growth hormone therapy can be a cure-all. But in rare cases, as with older people who chase the fountain of youth through growth hormone, the end of the story for these kids can be tragic.

A few years back, I saw a seven-year-old boy for loud snoring, nasal congestion, and poor appetite. He had been short and thin and had "fallen off the growth curve," a sign of growth delay. Placed on growth hormone, he had begun to grow. When I saw him, he was thin, but healthy. He also had enormous adenoids, an area of lymphoid tissue behind the nose that can grow in any child, but we see it a bit more commonly in kids taking growth hormone (growth hormone makes *everything* grow, not just the long bones). Since his snoring and congestion were significant, and his adenoids were enormous, his parents and I agreed on an adenoidectomy. The recovery from the operation is pretty easy, and he sailed through. But his dad noted that even weeks later, he wasn't eating well and wasn't growing as much, despite having been on growth hormone therapy for the past year. He also began complaining of stomachaches. His appetite diminished.

He went to see a gastroenterologist and had an upper endoscopy to look for reflux, stomach issues, esophagitis, and celiac disease. All looked normal.

Then his medical team ordered a CT scan of the belly, which showed his abdomen was riddled with tumors. He had widely metastatic desmoplastic tumors, involving the intestines, liver, and lining of the abdomen. He went to the top tertiary cancer-treatment centers, where patients with the most complex illnesses are seen, often referred from other cancer centers, and underwent multiple aggressive surgeries and adjuvant therapies. Despite all efforts, he suffered miserably and passed away of widely disseminated disease two years later. The death of a child is devastating. This is not to say that all growth hormone therapy causes such malignancies, but it is a real risk that many families consider before a child begins therapy.[17]

Adults, however, are more inclined to take HGH to curtail aging. It will not add height, but it is marketed to take off years of ever-present aging. Just as menopause ages a woman's body, primarily due to a decrease in hormonal levels, we can think of the natural decline in one's growth hormone (GH) levels as leading to overall bodily aging, termed by one group *somatopause*.[18] As humans, and all other mammals, age, the body's levels of GH decline, leading to some unwelcome changes such as decreased muscle mass, increased fat deposition, and decreased libido and energy levels, in both women and men. A landmark study in 1990, published in *The New England Journal of Medicine,* demonstrated that treating men over age sixty with GH for six months led to increased muscle mass, decreased adiposity, or fat deposits, increased bone density, and overall improved general well-being.[19] This study raised the interest in utilizing GH therapy as a method of extending youthfulness. Further studies looking at GH treatment in laboratory animals showed mixed results, including instances of speeding of aging and shortening of life span. It was later found that while giving GH to aging adults may reduce some undesirable features of aging, it may also increase growth of cancerous cells. The decline in GH with age may actually protect against malignancies.[20]

Many men combine HGH with testosterone supplements. HGH is not FDA approved as an antiaging remedy, but despite this, it continues to be used in injectable, pill, or spray forms. Claims are that it reduces fat, adds muscle mass, restores hair growth, boosts the immune system, and even improves memory. The Federal Trade Commission does not endorse any of these claims, nor does the FDA. If anything, diabetes and cancers similar to the horrific tumors of the child I described are real risks of HGH therapy for adults.[21]

VAMPIRE BLOOD AND STEM CELLS

Ever hear the expression "I have the heart of a twenty-year-old" from someone in his or her sixties? Well, the father of a close friend of mine said that and meant it. He had his first heart attack in his early forties, followed by several more, subsequent valve disease, and chronic heart failure. He was on disability by his midforties, with three young daughters and a wife at home. Fortunately he got on a heart-transplant list and was called at home late one night. An eighteen-year-old had been in a tragic motorcycle accident, which left him brain-dead. He was an organ donor and a perfect match for my friend's father. That night, he received his heart transplant, and the surgery was a success. He was sixty-two then, with the heart of an eighteen-year-old. He received a heart with no coronary artery disease, no scars, perfect valves, and strong cardiac muscles. Now, at eighty-five, he still has this young, perfect heart.

Organ transplantation continues to evolve, sometimes involving multiple organs (lungs, heart, liver, small bowel, pancreas, digits, limbs, uterus, face). Many success stories make the headlines, but few know what this costs—not just financially, but to the patient for long-term care, close monitoring of medications to suppress the immune system to maintain the organ, and continued risk of organ rejection. That said, receiving a young, healthy organ does, indeed, set the clock back, adding precious healthy years to one's life. What about other bodily substances? Can you receive blood from a twenty-one-year-old and set your clock back? The company Ambrosia, based in Northern California, claims that receiving a plasma transfusion from a young, healthy person will rejuvenate old, tattered blood. People pay close to $10,000 per infusion. At this writing, no data thus far shows that this provides vitality. Yet. Elysium Health claims that a pill containing molecular substances acting at the cellular level can add energy and years to one's life. They offer no evidence for this, either, although it's only $50 per month. The ingredients remain proprietary. Old brain? Take Nootrobox, which is claimed to enhance memory, clarity, and energy. The ingredients are similar to those in many caffeinated beverages and energy drinks, so they probably have some (transient) effect, but likely nothing permanent.[22]

Stem cells have become the latest and greatest over the past few decades, both in the medical and the lay communities. While stem cell therapy, especially with cells derived from embryos, has generated heated controversy of late, stem cell research actually began close to sixty years ago. A stem cell is like an unformed baby. It can become anything it wants, depending on multiple factors, both genetic and environmental. This ability to become any type of cell enables work with stem cells to move in any direction. Adult stem cells derived from bone marrow can form any type of blood cell in the body. This information stimulated the onset of stem cell transplantation for patients with severe hematologic (blood-related) cancers such as leukemia. Stem cell transplantation allows the recipient to start from scratch, building new, healthy blood cells after the body is rid of the cancerous ones. Simple, huh? Stem cell transplantation is not for the frail. It requires completely knocking out the entire immune system, ridding the bone marrow of any cell-forming capability, by giving extremely potent chemotherapy. The recipient is quarantined for at least a month, during which time he or she has no immune system. The stem cell transplant is then performed, with intravenous infusions over several weeks. The hope is that the transplant will work, with the recipient's bone marrow building a new set of healthy cells. This can take several months. Sometimes it does not work.[23]

The discovery that stem cells can treat illness by creating new, healthier cells has led to the building of new tissues—cartilage stem cells can generate new cartilage for ear deformities and joint injuries. Cardiac stem cells can generate healthy cardiac muscle; skeletal-muscle stem cells can repair muscle injury. As with any new innovation, the lifestyle industry became interested in using stem cells for rejuvenation of aging cells—skin cells for the aging face, cartilage cells for orthopedics, and neural cells for neurological disorders. Stem cell centers are now in every major city.[24] Skin stem cells have been used for wound healing and skin rejuvenation in cosmetic surgery.[25]

While there is no way to stop the clock or turn it back, we can slow it down. Appearance drugs, be they creams, lotions, injections, implants, or treatments with lasers and lights, will give the look of youth, but the core won't change, and will eventually sneak through to the surface. The oftentimes unmeasurable positive feeling of looking good is real, though, and should not be minimized. Remember, the placebo effect is real. A good haircut and highlights, a new outfit, or even a dermatological procedure can work wonders for one's sense of self, confidence, and well-being. These good feelings can carry one through until the next haircut, outfit, or treatment. The slippery slopes of these external changes can define who one is and become the one and only way to achieve a sense of youth and vibrancy. Drugs such as hormone replacements can be helpful, and many are safe when monitored closely by a trained physician. But taking these can be even slipperier than cutaneous changes, as they can do a dangerous number on and irreparable damage to your body.

Acceptance of aging is oft overlooked, but aging gracefully remains one of the most noble methods of looking (and feeling) younger. As with most issues regarding health, it always comes back to common sense—keep your brain young by continuing to learn, eat fruits and vegetables (the real stuff, not from a blender), and think positively about aging. None of us wants to suffer the ills that come with age, but it sure beats the alternative.

HYPE ALERT

* Many topical products can aid the illusion of looking younger (until you stop using them). It's one thing to look ten years younger and quite another to actually be ten years younger physiologically.

* Some antiaging therapies, such as testosterone and growth hormone supplementation, can be downright dangerous. You might feel younger, but you'll be at a greater risk for diseases such as heart attack, stroke, and cancer. In some instances, however, hormone therapy is beneficial; one example is children with growth challenges.

* Women should use hormone replacement therapy for the shortest time possible (primarily for symptoms of menopause). The decision to use it should be made case by case (as with so many other things in medicine). As for bioidentical hormones, there is no credible evidence that these work, nor are they regulated by any outside agency, including the FDA.

* The future holds much promise for extending life and slowing down the clock. Stem cell therapy, for example, is in its infancy. But for now, sometimes the old-fashioned tricks of eating well and avoiding smoking will do you well (and young).

13

HYPED EXERCISE: CLIMB EVERY MOUNTAIN

When Walking Beats Running

> WHAT DOES VIGOROUS PHYSICAL ACTIVITY DO TO THE BODY?
>
> CAN THE RISKS OF CERTAIN EXERCISES BE WORSE THAN BEING SEDENTARY?
>
> WHY IS SITTING CALLED "THE NEW SMOKING"?
>
> WHAT ARE THE "BEST" TYPES OF EXERCISE?
>
> CAN LOGGING TEN THOUSAND STEPS A DAY REALLY HAVE AN IMPACT ON YOUR HEALTH?

Many of us love extremes. We're either living a totally sedentary life on the couch eating cookies or training for marathons and buying wholly organic food. We know the old dictum about moderation, but it's hard to follow. In the past two decades, as the number of over-weight and obese individuals climbed, so did the number of people going to extremes in their fitness routines. Organized endurance-sport competitions—from classic marathons and triathlons to obstacle-course racing and ultramarathons—are more popular than ever. But do they promote a dangerous lifestyle? When is too much exercise just as bad as none at all?

The idea that too much of anything can kill extends to pretty much everything in life. Even water, which makes up about 66 percent of the human body, can turn toxic when consumed to such excess that it changes the chemistry of the blood. So as counterintuitive as it may sound, the same can be said for exercise. (As irony would have it, water intoxication is most common among athletes, as they attempt to rehydrate after an event. A 2005 study in *The New England Journal of Medicine* found that close to one-sixth of marathon runners develop some degree of water intoxication.[1] This can be lethal.)

Physical activity is important. As little as fifteen minutes per day confers substantial health benefits; forty to forty-five minutes five to six days a week is ideal, slashing your risk almost in half for premature death, diabetes, Alzheimer's, heart attack, and depression. People who routinely exercise have notably lower rates of disability, sleep better at night, tend to practice other healthy habits, and live on average about seven years longer than folks who don't ever break a sweat. But when does fitness become a liability? Let's explore this question among others.

You will be surprised to learn that to reap the benefits of exercise, you don't have to do nearly as much as you probably think. And in many cases, walking is better than running.

LET'S HEAR IT FOR THE AMISH

The Amish population, living primarily in states such as Pennsylvania and Ohio, and in parts of Ontario, Canada, has been a subject of interest for exercise specialists, dietitians, and geneticists.[2] This group emigrated to parts of the northeastern United States and Canada in the 1700s and, to this day, choose to remain removed from increasingly modern society. Their lifestyle, in many ways, mimics that of a typical farming community from over 150 years ago, and for this reason as well as their genetically close-knit community, the Amish have been studied by many groups to assess their unique lifestyle's effect on their health. Casting their higher rates of certain genetic disorders aside, largely due to inbreeding, I bring up the Amish because they have long been known to have much lower rates of the illnesses that plague the Western world, such as cancer. Before the emergence of increasingly more sophisticated fitness trackers, the concept of ten thousand steps per day was a notable feature of the Amish lifestyle. One study showed that Amish men walked, on average, eighteen thousand steps daily, not on a StairMaster or at the gym, but as part of a regular day's work in the fields. Amish women clocked, on average, fourteen thousand steps per day. While the Amish are not, by any means, marathon runners, their incidence of being overweight or obese is substantially lower than that of the general population. They don't eat what we'd call health food—they subsist on their traditional meat and potatoes, butter, lots of bread, eggs, and milk. They eat some vegetables, but are not found snacking on kale chips or celery stalks.

Even prior to the Amish studies of close to twenty years ago, Japanese walking clubs in the 1960s and 1970s promoted the idea of ten thousand steps per day with low-tech pedometers.[3] This is an auspicious number in Japanese culture, but it is arbitrary, not an actual data-driven marker for healthy levels of activity for both children and adults. It is likely an underestimation for the activity that kids should really have. For boys it's closer to thirteen thousand, and for girls, eleven thousand.[4] Despite more and more technological advances in wearable technology, childhood obesity and sedentary lifestyles continue to rise. Seventeen percent of our youth are obese, and in some communities, such as among Hispanics and African-Americans, that percentage hovers around 20 percent.

Health and fitness trackers are a nearly $1 billion business. By 2019, it's expected that close to 100 million devices will be sold annually worldwide. Wearing a health tracker does not equal having a healthy lifestyle. In fact, wearing one for the sake of losing weight has been found to have the opposite effect. A recent study looking at close to five hundred adults over eighteen months, all of whom were motivated to become more active and lose weight, found that those who wore fitness trackers lost about half as much weight as the nonwearing group. Those wearing trackers were also found to be less fit, as measured by degree of daily physical activity.[5] Another issue is that step trackers do not take into account degree of aerobic activity. While one does not need to be doing hard-core hill training or power sprints, brisk activity is important to maintain cardiovascular health, especially for kids. An often-overlooked issue is why the wearer is using the device. If it's purely for the fun of seeing the data, some studies have shown that the fun, and therefore the activity, declines with data checking.[6] The accuracy of data such as number of steps and heart rate also varies widely, diminishing the value of the trackers, with some overestimating step count by up to 15 percent.[7]

But for those who engage in little or no physical activity, the trackers can be motivational. The devices may also give the users some surprising, albeit undesired, information. I have an active friend who won't wear her fitness tracker at night because it depresses her to see how poorly she sleeps, and how little rest she's getting (yes, the devices can track sleep, too, which is an ingredient in overall fitness). I don't own a fitness tracker, but I gave one a go on a busy surgical day, when I do lots of surgeries, walk back and forth to the waiting room at least eight or ten times, walk back and forth to the recovery room all day, climb up and down six flights of stairs, and feel as if I'm moving all day. Nurses have jokingly offered to buy me a pair of roller skates. Sadly, I clock just a few hundred steps in an eight-hour day. No, thanks. Certainly not motivational data to me. My son wore a kid-version fitness tracker to sleep one night, and being a typically active nine-year-old sleeper, he clocked about three hundred steps. Without sleepwalking. My ultra-marathon friend has no use for a fitness tracker. (She just needs to cut her mileage a bit!) Her training runs alone would probably drain the tracker's battery in one afternoon.

Insurance companies are now offering free fitness trackers with policies for health, life, and long-term care. They provide so-called financial incentives if you not only wear the fitness tracker, but provide personal data that the company can upload at will. This is not just data for steps, sleep, and heart rate—it's information about where you shop for groceries, which groceries you buy, your online buying trends, and where you travel. In exchange for this private information, you receive insurance discounts and coupons, the latter of which are clearly tied to your now-shared buying habits.[8] Personally, I don't want strangers— and the internet—knowing I like to buy Snickers bars and travel to the desert even if these details will make my insurance policies more or less expensive. But that's another story.

YOUNG AND OLD COUCH POTATOES

While adults are oftentimes better off going for a twenty- to forty-minute brisk walk several days per week, kids need to move to the point of working up a bit of a sweat. Not all day every day, but part of the day every day—for at least thirty to sixty minutes. With more and more schools reducing outdoor recess time and physical education classes, and piling on more and more nightly homework, kids are losing precious minutes of exercise daily. And it shows. Over 24 million children in this country are obese or overweight, and the number continues to trend upward. Only one in four children get the recommended daily exercise (thirty to sixty minutes of moderate to vigorous activity). The life expectancy of these kids is shorter than that of their parents, and we're seeing earlier and earlier diagnoses of type 2 diabetes, obesity-related chronic illness such as fatty liver, exercise intolerance, and obstructive sleep apnea. Exercise intolerance is the inability, because one is so out of shape, to perform much exercise at all that entails physical exertion and breathing hard. The lungs and heart simply aren't fit, and severe postexercise pain, fatigue, nausea, and other negative effects happen. Kids as young as eight years old are being considered for gastric bypass surgery.[9] Most pediatric tertiary-care centers now have a designated department for obesity-related surgery.

But, on the other end of the exercise spectrum in children, some kids, also as young as eight, suffer from overuse sports injuries as a result of extreme pressures to excel early.[10] Kids are specializing earlier and earlier, choosing one sport to focus on, and doing that sport alone, every day, for up to four hours, beginning as young as age four or five. For some sports, such as golf, tennis, figure skating, and gymnastics, this may be the only path to success. Look at pictures of golf legend Tiger Woods, barely out of diapers, on the putting green. Or Andre Agassi, former tennis great, hitting the courts as a toddler. But many of the team sports, including football, soccer, basketball, swimming, and hockey, to name a few, are getting more and more competitive, and parents, primarily, want to get any edge they can for their preschooler to get a sports scholarship to college. With this shift, gone are the days of "everyone gets to play." Alas, the days of "everyone gets a trophy" still remain. We are seeing a substantial rise in stress fractures, joint disorders, plantar fasciitis, and muscle injuries with the increasing focus on single sports by young athletes. Not to mention increased incidence of concussions and long-term effects on brain development.

Early on, sport diversification should be encouraged. We want the next generation to be active and healthy. But we also want to keep them out of the orthopedic and neurosurgical operating rooms before they hit double digits. We also don't want previously menstruating adolescent girls to stop getting their periods for months due to inadequate body-fat stores and low hormone levels, all from athletic pressures.

So where is the sweet spot for daily activity? Millions of kids are living a sedentary lifestyle never seen in generations past. Thousands of kids are getting pushed, by parents, coaches, and themselves, to be mini superathletes, only to suffer injuries due to overuse or downright trauma. We can blame a great deal of the sedentary lifestyle in kids on use of technology, with some kids on their devices for over eight hours each day, and well into the night. We can also blame parents for allowing this easy access too often and too early. For a while, the American Academy of Pediatrics advocated absolutely no screen time (including television) for children under age two. Once the Baby Einstein fad was found not to create baby Einsteins, this became easier to follow. But once the second birthday hit, there came limitless screen time for all. Not good. We can also blame not food per se, but ease of access to food and food's convenient portability. Kids do not need a snack the second they finish running around for an hour. But as Pavlov would have it, many kids equate a team practice, a game, or a trip in the car home from school with snack time. And it's been made easier—even foods that used to require a utensil are now in suckable pouches.

Couch potato adults abound, many of whom, as is the case with their kids, are tethered to their screens. Hours can pass in what seems like the blink of an eye when scrolling through social media pages, news sites, or emails. On the flip side, super-body-obsessed gym rats, who may be doing more harm than good with that extra mile, extra fitness class, or extra round of weights, abound. As with vitamins or healthy food, more does not necessarily equal better.[11] Moderation is key.

To some of us, exercise is like a drug. As with anyone addicted to a drug, we need not the high from doing it, but to avoid the low we feel without it. "It's like brushing my teeth," a friend of mine into athletics has told me. "It's not that it's so great, but I feel pretty crummy without it." Other friends of mine who have run multiple marathons are adrenaline junkies. They frequently sign up for races of all kinds. It's not the win they're looking for—it's the experience, the crowd, the scene. I completely relate to all of this. A friend who always found my choice to run in crowds a bit odd calls it "the lemming effect."

I follow a strict regimen of running five mornings per week. I clock about twenty to twenty-five miles each week. But some days as I'm driving home from work and see a runner on the street, I remark to myself that I'd love to go for a run that day, completely forgetting that I had run that morning. I don't give it a thought anymore. Indeed, it's like brushing my teeth. I don't feel particularly great about it, but I do feel crummy without it. So what do we toothbrushers do to feel the effects of exercise? We do more. Push it a little bit and you'll remember your run from that morning. Forget brushing your teeth—it's a trip to the dentist, laughing gas and all. This is where those who regularly exercise ramp it up a notch—if a five-mile run is good for you, a ten-mile run is better, and a fifteen-mile run is even better than that. But this is not the case; there is a point of diminishing returns. This goes not only for skeletal muscles and joints, but also for another important muscle—the heart.

Most of us are aware of the general benefits of exercise (even when we avoid it) when performed in the right amount. It has positive effects on virtually every system in the body—helping us to keep our weight managed, our metabolism humming, our heart healthy, and our brain sharp. But exercise at "ultravigorous" levels for prolonged periods causes something else to happen. Look no further than marathon runners to understand this law of diminishing returns. When the heart is stressed, most notably during a heart attack, troponin is released. When patients are admitted to a hospital for a possible heart attack, we measure the level of troponin in their blood as a marker for the extent of heart-muscle damage. Up to one-half of marathon runners display high levels of troponin, even as high as in a heart attack patient. Their hearts have been stressed, even without typical symptoms of chest pain. Just as a heart attack can leave a scar, so can regular damage from high levels of troponin.

High-endurance exercise such as marathons and triathlons can also do a number on the kidneys. During extreme exercise or even from acute muscle injury, skeletal muscles can begin to break down, literally digested, by enzymes, undergoing rhabdomyolysis. The protein by-products from this go to the kidneys, leading to a decrease in excretion and then later to blood in the urine. Almost all marathoners will have some trace blood in the urine, from rhabdomyolysis. Some routinely pee out bright red blood after a race, despite adequate pre-race and intra-race hydration. Even "spin class–induced rhabdomyolysis" is now cropping up in the medical literature.[12] In 2017, the media reported on a thirty-three-year-old schoolteacher in the Bronx who, after fifteen minutes of intense pedaling on a bike, felt nauseated and nearly passed out. Granted, she was a novice at the sport, but you wouldn't think an otherwise healthy young woman would meet such a dire diagnosis. Dr. Maureen Brogan, a kidney specialist at Westchester Medical Center, where this woman was treated, detailed this case in a report. In two years she and her colleagues saw six cases of spinning-related rhabdomyolysis. One woman, also thirty-three, had kidney failure and needed dialysis for a month until the organs recovered. These cases are rare, but they do point out the perils of overly intense exercise—even when you are young and healthy.

CAN YOU OM YOUR WAY TO FIT?

The practice of yoga dates back anywhere from five thousand to ten thousand years. Because much of the early practice and teachings were oral, or written on delicate leaves, the dates are a bit unclear. It began as ritual songs, texts, and mantras, having nothing to do with downward dogs or child poses. Centuries after its initial years, yoga masters created tantra yoga, with techniques to cleanse the mind and the body to explore the connection between physical and spiritual existence. And so began the Westernization of yoga, or hatha yoga. Now, yoga centers are on every corner, some right above a Starbucks or a yogurt shop. As with so many health practices, it first landed in the United States in 1947 in the town of, you guessed it, Hollywood.

Yoga remains one of the most well-balanced forms of physical activity, combining balance, stamina, and flexibility with breath control, relaxation, and strength.[13] My kids did yoga as part of their preschool curriculum. We did some of the exercises at night as a family, which seemed to get everyone a bit calmer before bedtime. Parents included. Yoga classes require focus, patience, and silence. I have been tied to a hospital pager since 1991, so I have never tried it. I do envy yogis. It's just not for me. But hot yoga is another animal. If yoga is good, hot yoga must be better, right? Yoga evolved over thousands of years and continues to do so. Hot yoga is yoga performed in a room where the temperature is cranked up to an average of 105 degrees. Just a few minutes after the first few poses, instructor and students are dripping with sweat. The notion is that, by sweating more, it somehow makes for a better workout than doing the same class in a room with a humane temperature. Also, the idea is that sweating will release all of those nasty toxins more effectively if it's more vigorous and speedy, with the help of a hotter room temperature. Sorry, folks. Neither is the case.

While sweating is a fine function of the excretory system, along with urinating, defecating, and salivating, it does not burn calories nor does it leech out those evil humours intoxicating our bloodstreams. Sweat consists primarily of water, along with other bodily wastes, including salt, sugar, urea, and ammonia. The latter two are by-products of protein breakdown. Sweat does not contain BPA, pesticides, asbestos, pollutants, or other ills we wish to flush out of our system. While it's certainly good to sweat, all of the contents of sweat are continually being replenished—most notably water, which you should definitely drink in high quantities after a hot yoga class. But again, sweating is not equivalent to burning calories. While one does burn more calories in warm weather than cold weather (unless while working out you're shivering, which is a calorie burner), hot yoga does not burn more calories than regular yoga comparing class to class. So, if you want to gain the benefits of yoga, which is great for strength training, flexibility, mindfulness, and even cardiovascular fitness, go for it. But don't pay extra to sweat extra.

AVOID GET-THIN-QUICK FIXES AND SITTING DISEASE

As the saying goes, if it's too good to be true, it probably isn't. This certainly holds for any weight-loss or get-fit-quick gizmo. The healthiest people, defined by good diet, good habits, and routine exercise, know this best. Getting in shape takes time, discipline, and continued work. One friend of mine has to nearly lock herself in her house (away from her free weights) so as not to exercise, as if a few days off will turn her body to mush. Days when she has bronchitis. And it's raining. While this is one extreme, habits do lead to the best results. There are no quick fixes. Although most of us know this, over $30 billion a year is dropped in the United States alone on weight-loss products, which include not only the widely criticized (and rightly so) supplements that claim to rev your metabolism and burn your fat, but also get-fit-quick products. Some of the most well-known of these were touted by then beach babe Suzanne Somers, famous as the hit TV show *Three's Company* hottie of the 1980s. Who wouldn't want buns and thighs like hers? Thus came "buns of steel" and Thighmaster. Unfortunately, these products don't come with Suzanne's thighs or butt, and no matter how hard one tries, these products will not deliver—certainly not as stand-alone fixes. But at least these products required that the user do something. Products such as sauna suits, vibrating platforms, any weight-loss pill, spandex tights, or toning shoes do nothing. Nor do balance bracelets, which claimed to improve physical prowess and became popular with kids as well as adults in the mid-2000s. One product even added weights to utensils, with the goal of adding some arm work while you shovel in that dessert.[14] No such luck.

While gimmicks such as wearable technology, clothes to sweat in, or fitness toys may not be the be-all and end-all to achieving fitness goals, they are not harmful, except maybe to your wallet. What's actually detrimental to fitness is no fitness at all. Sitting, that is. Even for those who exercise daily, sitting throughout the day can be downright dangerous to your health. While this seems intuitive, with a metaphorical couch potato coming to mind, or someone who looks like Homer Simpson, sitting for long periods may be harmful even if you're clocking in ten thousand steps, thirty to forty-five minutes of moderately vigorous activity, and eating and sleeping well. A recent study looking at forty-seven other studies assessed sedentary behaviors in adults and their association with mortality and several illnesses.[15] Each study found that, despite regular exercise, all-cause mortality, cardiovascular-related mortality, cardiovascular disease, cancer-related mortality, cancer diagnosis, and diabetes all increased with an increase in sedentary behaviors. With long-term sitting, minute-to-minute calorie expenditure declines, effectiveness of insulin declines, leading to increased risk of diabetes, and even bad-cholesterol levels rise, leading to increased risk for cardiovascular disease. Given this, many fitness trackers now have a buzzer that reminds the wearer to get up a bit and walk around if he or she has been sitting too long. While data is lacking on whether this will reduce sedentary-related morbidity and mortality, it certainly can't hurt. According the World Health Organization, physical inactivity is the fourth-leading worldwide risk of mortality, thought to be responsible for about 3.2 million deaths annually.[16]

Clearly, sitting all day is not good, and some have even called sitting the new smoking, as it seems to confer risk profiles similar to those seen in long-term smokers. If recommendations to get up every now and then, either by self-reminders or by a buzzer on one's wrist, don't work, some think maybe we should just stand. Thus was born the standing desk.

These high-legged worktables resemble worktables of artists and architects, who prop themselves on high stools for hours on end. But these desks don't come with stools. You just stand. The idea is to burn more calories per minute, lower your cholesterol, and improve glucose metabolism. In many ways, the standing desk has been a great answer for those who need to work for prolonged periods at a desk.[17] We surgeons stand all day; one might say that we stand in one place all day. Trade the word *desk* for *operating table,* and there's your standing desk. We've been doing this for years.

While some surgeries are performed while the surgeon sits, for instance those done under a microscope, through a telescope, or on the eyeball, most are performed with the surgeon standing. The surgeon's standing desk, the operating table, has been around for centuries, long before the new trend for standing desks was popularized. We never considered standing at an operating table for hours on end to be a fat-burning endeavor. The known liabilities of standing at a desk (i.e., operating table) are not new to us. Most surgeons, especially those who are tall and lanky, suffer some back pain—upper and lower, primarily. Upper due to a cranked neck in a locked position while dissecting nerves or blood vessels; lower due to increased lumbar pressure from standing all day. Many of us wear clogs in the operating room. While some of us have stylish ones (I have pink-flowered ones for the spring, snowflake ones for winter, as well as several other decorative choices), we wear these because they are the most comfortable shoes in which to stand still for long periods, easing lower-back pain. Standing at the operating table for hours every day puts us at risk for nasty varicose veins or, worse, blood clots. Many surgeons put on compression hose first thing every morning, to minimize this occupational hazard. As noted earlier, male surgeons, especially those who perform longer surgeries, have an extremely high incidence of kidney stones from standing for long periods and being routinely dehydrated. All of these issues came to the surface with the rage of standing desks, but the issues are not new. And as is recommended to surgeons, it's best to have the desk at a comfortable height, and it's best to take breaks, by either walking or sitting, throughout the day. Again, moderation wins.

THE BEST IS WHAT'S RIGHT FOR YOU

So what is the best way to get into great shape and have a svelte physique? No single thing will do the trick for everybody, and neither will one single thing for *you*. You have to find what works for you in the exercise realm, so long as you do what you love, and mix it up: part cardio, part strength training, part flexibility. Just as there is no single best diet, there is no proven recipe or formula for permanent weight loss and total fitness. If fitness trackers don't work for you, don't feel guilty. Find other motivators. When the trends don't match your preferences, that's okay.

One of my main goals in writing this book is to impress upon you the power of individual choice even in matters of health and medicine. When patients ask me what's the best surgical technique I'd recommend for a given surgery, my answer is always the same: "The best technique is the technique that your surgeon does best. And technology is no substitute for technique."

Such a narrow example can expand to all issues related to your health. There is no best exercise, best gizmo, best diet, best supplement, best diagnostic test, best drink, best skin cream, or best medical decision. There is no perfect genetic makeup. Every road one takes for health has multiple forks. With these forks, decisions are made and directions change, hopefully for the better. As we continue to learn, we continue to modify what we think is ideal for health, and not. Projects investigating the power of the human microbiome, for instance, are bound to be game changers. But don't ask me today which friendly strains of bacteria to look for in your probiotics that will work with *your* microbiome. Those answers are still being figured out.

What was your best exercise regimen in your twenties may not work for you in your sixties. That annual mammogram that was such a standard practice in the past may not be the best way to go in years to come. The field of personalized medicine and health is nascent, but perhaps someday we will have answers to what is best for each of us at any time of our life. The exciting thing about our health is that we are continually learning about what the human body is capable of, and what is just impossible to withstand. Recommendations and guidelines of today will likely be disputed in years to come. If we recognize the whirl-wind of change at every turn, I hope we will be less inclined to grasp on to the latest and greatest of just about anything, especially when it comes to our health. Because sooner than we think, it will be history.

HYPE ALERT

* It doesn't take much exercise to reap its benefits. For many people, walking throughout the day and putting a little bit of effort into it will do you good.

* More exercise is not always better. At the extreme end, with vigorous exercise for extended periods, the risks outweigh the benefits.

* Kids needs to get in their activity, and sports diversification is important. This helps them learn new skills and prevents injuries.

* Sweating does not necessarily burn calories. So hot yoga does not burn more calories than regular yoga when compared class to class.

* Fitness trackers and products marketed to slim you down and shape you up quickly (think Saturday-morning infomercials) rarely deliver as promised, but rarely will they harm you. They can be a gateway, however, to other forms of exercise that do you good.

* Try not to sit for prolonged periods. It may be as bad for you as smoking.

* Just as there is no best diet, there is no best exercise. Do what's right—and fun—for you (no hype included).

DON'T BELIEVE THE HYPE

Is It All Hype?

In 2017, after a 146-year run, Ringling Bros. and Barnum & Bailey Circus took down its tents. Many of us have fond childhood memories of it, or memories of our children or our children's children marveling at The Greatest Show on Earth. But for many reasons, only one of which was concern for maltreatment of animal and human employees, the century-and-a-half-old icon closed for good.[1] But awareness of animal rights and cruelty to both animals and humans is not new. And the thrill of the extreme, the bizarre, or even the impossible didn't start, nor will it end, under a circus tent. P. T. Barnum was a notorious master of the hoax, not to mention of abuse, and to this day circus-goers have continued to crave the extreme, the strange, the impossible. The trick. The hype. Even in the years before the renowned circus came to be, Barnum would advertise the exhibit of a beautiful mermaid, which turned out to be an ape's head sewn to a fish's body, or the claim that George Washington's 161-year-old nurse would be on-site to tell tales of our dearly departed first president.[2] Goodness—weren't people gullible back then? A mermaid? A 161-year-old woman? Well, current hoaxes may not be as rudimentary and crude, but thanks to advanced technology and social media, they are certainly more widespread. And just as ridiculous, if not more so. And animals remain in the picture.

In 2014, Jimmy Kimmel and his team launched the "Sochi Wolf" prank, posting videos of a feral wolf wandering the halls of U.S. athletes' dormitories at the Sochi Olympics. The videos were actually shot in a Los Angeles studio, and the wolf was a rescue animal found in North America, not Russia. But close to 6 million viewers fell for it on YouTube.[3] A wolf in an athletes' dormitory? Goodness, we *are* gullible.

While hoaxes such as circus tricks and viral YouTube videos of wild animals can make us cringe, feel duped, or even laugh at ourselves for believing the unbelievable, we do not take it well when we're duped about health advice. When information bombards us hourly, who's to know what is fact, fiction, or somewhere in between? Millions of us now sport personal fitness trackers, which may provide not just information from output (steps, calories burned, active minutes, heart rate, and the like), but also input (calories in, fluid intake, hours and quality of sleep). This personal apparatus on our wrist is about us and us alone—not a recommendation from a fitness magazine, Twitter guru, or a doctor. It's personal. Unique. This uniqueness is what we crave, and it is likely the future of medicine. In accuracy, on the other hand, the gray zone prevails. Did I really only walk 8,957 steps today? My heart rate only climbed to 109 beats per minute? I know I slept more than six hours and fourteen minutes. Didn't I?

This technology is still nascent, and accuracy is not yet the norm. But as personal health technology, or "smart medicine," surpasses the smartphone, not only will accuracy improve, but so will the sophistication of personal monitoring and testing. Not only can heart rate be traced, but placing one's fingers on a sensor can transmit one's heart rhythm, similar to a traditional electrocardiogram, directly to your doctor, who might well be sitting in front of a screen as if you were in the exam room anyway. Some phones will be equipped with ultrasound probes, enabling a veritable picture to be sent to a doctor when you have an ache or a pain. Personal sensors can also calibrate and transmit vital information such as blood pressure, body temperature, and respiratory rate. And this is only the beginning.[4] Software for personal genome mapping is already in the works. All of this is unique, personal information—*your information*—but for personal or public access?

Having personal health information at the touch of a button may come with a price—the increased chance of a privacy breach in the increasing complex of Web-based personal data. Large health insurance companies have had hundreds of thousands of Social Security numbers hacked; large consumer companies have had hundreds of thousands of credit card numbers hacked. If we want easy access to our genetic makeup, can that be hacked, too?

Even if one is not sporting a tracking device or an ultrasound probe, the ability to search for information is almost *too* easy and can leave the searcher more baffled after a good Google search than before. A friend of mine had intermittent abdominal pain and had six full days to consult with the good Dr. Google before seeing her own doctor. And perhaps sometimes (sadly) Dr. Google can be more straightforward than the real thing. As in any profession, medicine is filled with fearmongers and swindlers, preying on the fear of illness or simply the quest for health. While most physicians do have the patient's interests in mind and do care about doing the right thing, and doing no harm, some can dupe even the savviest patients. Many of the questionably ethical physicians will have posh offices, beautiful waiting rooms, the highest-quality websites, and top-of-the-line products for purchase with your visit. While these are not by themselves signs for you to raise an eyebrow, I recommend buyer beware when your health-care provider's office starts to feel too much like a spa.

One of the challenges of writing a book such as this one stems from our breakneck access to new information. An issue that seems as if it will be an evergreen headline, no less a critical link to our health, fades in the forest sooner than we'd think. Vitamin E was a must-have supplement to prevent cancer, until it was found to be a potential cancer cause. Low-fat anything was considered to be the answer to dieters' woes, until it was later discovered that sugar, not fat, may be at the root of the obesity epidemic.

The other challenge is a good one—we as humans crave the new and exciting, whether it's a mermaid, a former president's nurse, or a wolf in a hallway. But when it comes to our own health, we crave to know what's right—have the upper hand, be one step ahead—so as to not only live longer, but also to live better. One of the goals of this book was to provide you with the tools to think critically when reading, hearing, or seeing new health information. The knowledge of today is the fool's thought of the future. We doctors are often on the front line of discovery of health information, and one would hope we would be among the first to shoot down new things as soon as we hear that they have been proven false. Just recently, the American Heart Association turned a 180 on coconut oil. For several years, this venerated organization, focused unquestionably on heart health, recommended coconut oil over some other fats as "heart healthy." A newer study found this to be false, and the recommendation was switched, again with a headline. At this writing, the debate rages on over coconut oil's use. But it brings up a good point: even respected medical associations can flip-flop as new evidence emerges (and then switch again farther down the road). This is why it's important to keep an open mind as science evolves.

Doctors used to not only smoke cigarettes, but were advertisers of specific brands, touting their refreshing mint taste and cool, relaxing effects, without any notion that these would be the source of millions of premature deaths. Now doctors are the ones damning cigarettes. The medical field has remained much less adamant about alcohol—with so much wavering. Almost a century ago, we were in the throes of Prohibition, when alcohol, regardless of form, brand, or amount, was illegal without a doctor's prescription. Today the National Institutes of Health, the most reputable and competitive government agency to support medical research, is funding a single-blinded, prospective, randomized study to identify whether daily consumption of any alcoholic beverage is healthy or harmful.[5] And even more interesting, alcoholic beverage companies such as Anheuser-Busch are participating in the project.

While this concept screams "conflict of interest," the participating companies are fully aware that the study may conclude that alcohol is harmful. They are taking the high road on this one and will, if findings are negative, promote safe drinking practices. But would this mean abstinence? Would they take the risk of putting themselves out of business? And if the findings are positive, deeming alcohol consumption to be healthy, will there always be a question of conflict of interest, given that the study was financially supported by alcoholic beverage companies?

Medical studies, big or small, are often tricky to interpret. Here's a real-life example from my own practice. I have found a 100 percent correlation between toddlers who aspirate whole nuts—leading to choking episodes and trips to the operating room for life-and-death surgery and pediatric ICU stays—and their being unimmunized. Three for three—all choked on nuts; none were immunized. And all three had comfortably upper-middle-class, college-educated parents, and all had primary care doctors who condone healthy eating and "flexible" vaccine schedules. The American Academy of Pediatrics advises against eating whole nuts (due to choking risks) in children under age five and strongly advises against veering from the vaccine schedule recommended by it, the CDC, and the AAFP (American Academy of Family Physicians). So there's my case series! I can see the headlines now: "Delay in Vaccinations Causes Food Choking Accidents in Toddlers of Wealthy College-Educated Parents!" The nonsense in that is clear, but certainly this was a case series, with 100 percent correlation of all data points! Be careful and thoughtful when you read something that sounds incredible. It probably is.

ACKNOWLEDGMENTS

Whether you're like many of us and go straight to a book's acknowledgment pages before cracking the first chapter, or are reading this at its conclusion, you wouldn't be reading this if not for Kristin Loberg. While Kristin and I share an alma mater (Cornell! Go Big Red!), there's a decade between us, and it took her son Colin to introduce us just a few years back, three thousand miles from Cornell, here in Los Angeles. I am ever indebted to Kristin for hearing my ideas when this book was in its infancy, sticking with me through to this project's fruition, providing insights, stories, writing finesse, and so much more to make this book what it is. She can tell it like it is and says it the right way. She and I live in a world of hype, Hollywood, and herbs, and she has been a continual voice of wisdom, words, and support from start to finish. She is not only a writing collaborator, fellow running aficionado, and fellow juggler of work and home life, but also a friend. I thank Kristin's family—Lawrence, Colin, and Teddy—who no doubt have heard rumblings of this book during its circuitous progress. Years before I met Kristin, I met her incredible father, Eric, a fellow Cornellian turned UCLA professor, who welcomed me as Big Red family soon after I arrived to the West Coast. I thank Eric for his kindness, lust for life, and giving Kristin so much of the élan she has. He left us too soon, but his voice is heard in these pages.

Referring to Amy Rennert as my literary agent just doesn't seem right. Don't get me wrong—she *is* my literary agent and is an amazing one. But Amy is also a lifelong friend. No, she's family. We go back more decades than I'd like to reveal (okay, almost *five*). She took a chance on me several years back as a newbie author and has stuck by my side with heart through *Hype*. Her keen eye for hype is sharp, but she also knows when to rein it in and hear both sides. Both Amy and Louise bounce medical issues off me, gauging "hype factors," and have given me valuable insight into what people need to hear and know. And how they need to hear it.

Daniela Rapp at St. Martin's believed in this project from the get-go and has turned an idea into reality. She has applied her ability to see the larger picture of where our message is going, and how best to share what we need to say, with grace, style, and smarts. Her input has been invaluable, and her patience, encouragement, and no-nonsense view of health issues have provided critical additions to the book. She also happens to be an incredibly cool person who always wears a bike helmet. The team at St. Martin's Press, including Sarah Becks, Michelle Cashman, Laura Clark, Paul Hochman, Gabrielle Gantz, Rebecca Lang, and Lauren Jablonski, have helped carry this project to the finish line. I am ever indebted for their wise words, support, smarts, and enthusiasm. . . .

Megan Beatie has championed this book to get heard and has simultaneously become a friend and travel pal. She knows hype when she sees it, finds when and where it needs to be revealed, and creatively angles ways to get our message across. I am a lucky person to have Megan on my side.

I have so many people to thank, many of whom contributed ideas, their own stories, or their patients' stories and gave me reason to keep writing, knowing that hype abounds and needs to be bashed. There are too many to name, and names omitted are nobody's fault but mine, but some include Amy Eldon and Jon Turteltaub, Annaka and Sam Harris, Ellen Ferguson, Dave Price, Kathleen Lago, Bunny Ellerin, Kim Dower, Samantha Ettus, Jane Bendor, Bonnie Solow, Amy Sommer, Jana Gustman, Susanne Resnick, Joanna Moore and Brad Ramberg, Craig Underwood, Nicole Kennedy, Melinda Benton, and Louise Kollenbaum.

Fellow physicians are so key to my day-to-day world. Even if we don't share every opinion, we do share experiences, patients, and patience. Dr. William Silen set the stage for who I am as a doctor, and some of his trainees helped jar my memory of my year on his surgical team. Drs. Steve Teitelbaum, Jamie Watkins, Arthur Lauretano, Neil Bhattacharyya, and Mark Brown all shared "Silen rounds" with me, and we lived to tell the tale. Drs. Corinne Sadoski and Betsy Watson, my medical school roomies, lived with me through the earliest, and in some ways toughest, days of becoming a doc. They bash hype every day.

UCLA is a special place, and I thank the institution for the privilege to work, teach, and learn there for over two decades. Many of the folks at UCLA and in its environs had specific impacts on this book's stories. Some include Drs. Judith Brill, Swati Patel, Ihab Ayad, Wendy Ren, Alisha West, and the UCLA Head and Neck Team. Drs. Cara Natterson, Jody Lappin, Lisa Stern, and Farah Hekmat, who also had some fascinating perspectives to share. In addition to physicians, many other colleagues have also added so much to patient care and to the content in this book. Rachel Giacopuzzi-Brown, Nancy Villegas, Erik Phelps, Lara Ginnetti, Traci Kellum, and Dierra Merritt have weathered many storms with me.

Family is everything, providing the most support one could ask for. My father, Dr. Stanley Shapiro, and my brother, Dr. Adam Shapiro, are the first physicians I ever admired and learned from, and we continue to discuss issues and bounce medical banter around the Thanksgiving table. My mother, Dee Shapiro, has done all that she can to hold back from asking daily (or even twice daily) about the book's progress. But her ardent dedication, love, story ideas, perpetual hype-alert radar, and support of me are as powerful as one could hope for. And that's not just the placebo effect. My husband, Elliot, is more knowledgeable, well-read, and hype-free than just about anyone else on this planet. As a cancer surgeon in Los Angeles, he's heard and seen it all. He denounces pseudoscience, yet understands and even appreciates where it's coming from. He has nurtured and supported this project, providing new angles, references, and ideas, from the obvious to the esoteric. My kids, Alessandra and Charles, have taken it upon themselves to see through nonsense, be it health hype, social media hype, or otherwise. They and so many kids their age are our future and will hopefully read this as historical nonfiction and find (and bash) new hype in the years to come.

NOTES

The following is a list of books, scientific papers, and Web citations that you might find helpful in learning more about some of the ideas and concepts presented in this book. For updated information and access to new insights, visit www.drninashapiro.com.

1. A Site to Behold: The Wild West of Internet Medicine

1. Rick Nauert, "Why First Impressions Are Difficult to Change: Study," *Live Science,* January 19, 2011.
2. Chris Reid, "The Difference Between Search Engine Optimization and Search Engine Manipulation," *Constant Content,* November 9, 2016.
3. R. M. Merchant, K. G. Volpp, and D. A. Asch, "Learning by Listening—Improving Health Care in the Era of Yelp," *JAMA* 316, no. 23 (December 2016): 2483–84.
4. A. Milstein et al., "Improving the Safety of Health Care: The Leapfrog Initiative," *ResearchGate,* November 1999.
5. C. K. Christian et al., "The Leapfrog Volume Criteria May Fall Short in Identifying High-Quality Surgical Centers," *Annals of Surgery* 238, no. 4 (October 2003): 447–57.
6. www.healthwebnav.org.

2. Risky Business: What Ebola and Your Car Have in Common

1. https://www.cdc.gov/media/pressrel/r2K0107.htm.
2. https://www.cdc.gov/motorvehiclesafety/teen_drivers/teendrivers_factsheet.html.
3. https://www.skincancer.org/skin-cancer-information/skin-cancer-facts.
4. R. K. Masters et al., "The Impact of Obesity on U.S. Mortality Levels: The Importance of Age and Cohort Factors in Population Estimates," *American Journal of Public Health* 103, no. 10 (October 2013): 1895–1901.

5. http://who.int/csr/disease/ebola/en; and http://www.unaids.org/en/resources/fact-sheet.

6. https://www.cdc.gov/mmwr/preview/mmwrhtml/mm4909a1.htm.

7. http://www.cdc.gov/injury/wisqars/pdf/101cid_all_deaths_by_age_group_2010-a.pdf.

8. http://www.cdc.gov/reproductivehealth/maternalinfanthealth/infantmortality.htm.

9. http://www.cdc.gov/nchs/fastats/child-health.htm.

10. Ibid.

11. T. Roseboom, S. de Rooij, and R. Painter, "The Dutch Famine and Its Long-Term Consequences for Adult Health," *Early Human Development* 82, no. 8 (August 2006): 485–491.

12. http://blog.oxforddictionaries.com/2014/11/oxford-dictionaries-word-year-vape/.

13. http://blog.oup.com/2014/11/e-cigarette-vape-timeline/.

14. J. Brown, "Real-World Effectiveness of E-cigarettes When Used to Aid Smoking Cessation: A Cross-Sectional Study," *Addiction* 109, no. 9 (September 2014): 1531–40.

15. R. E. Bunnell et al., "Intentions to Smoke Cigarettes Among Never-Smoking U.S. Middle and High School Elective Cigarette Users: National Youth Tobacco Survey: 2011–2013," *Nicotine & Tobacco Research* 17, no. 2 (February 2015): 228–35.

16. E. O'Malley Olsen, R. A. Shults, and D. K. Eaton, "Texting While Driving and Other Risky Motor Vehicle Behaviors Among U.S. High School Students," *Pediatrics* 131, no. 6 (June 2013): e1708–e1715.

17. A. E. Carroll, "Alcohol or Marijuana? A Pediatrician Faces the Question," *New York Times,* March 16, 2015.

18. M. Cerda et al., "Medical Marijuana Laws in 50 States: Investigating the Relationship Between State Legalization of Medical Marijuana and Marijuana Use, Abuse, and Dependence," *Drug and Alcohol Dependence* 120, no. 1–3 (January 2012): 22–27.

19. G. Lopez and S. Frostensen, "How the Opioid Epidemic Became America's Worst Drug Crisis Ever, in 15 Maps and Charts," *Vox,* March 29, 2017.

20. J. M. Mullington et al., "Cardiovascular, Inflammatory and Metabolic Consequences of Sleep Deprivation," *Progress in Cardiovascular Diseases* 51, no. 4 (January–February 2009): 294–302.

21. https://www.cdc.gov/ncbddd/hearingloss/facts.html.

22. https://www.amazon.com/AncestryDNA-Genetic-Testing-DNA-Test/dp/B00TRLVKW0.

23. C. Seife, "23andMe Is Terrifying, but Not for the Reasons the FDA Thinks," *Scientific American,* November 27, 2013.

24. Harvard Women's Health Watch, "Direct-to-Consumer Genetic Testing Kits," Harvard Health Publishing, September 2010.

25. https://www.cdc.gov/vhf/ebola/outbreaks/history/chronology.html.

3. Turf Wars: An Important Lesson in Correlations

1. A. B. Hill, "The Environment and Disease: Association or Causation?," *Proceedings of the Royal Society of Medicine* 58 (January 14, 1965): 295–300.
2. V. V. Kumar, N. V. Kumar, and G. Isaacson, "Superstition and Post-Tonsillectomy Hemorrhage," *Laryngoscope* 114, no. 11 (November 2004): 2031–33; and R. F. Baugh et al., "Clinical Practice Guideline: Tonsillectomy in Children," *Otolaryngology—Head and Neck Surgery* 144, no. 1 (suppl) (January 2011): S1–S30.
3. C.A.M. Bondi et al., "Human and Environmental Toxicity of Sodium Lauryl Sulfate (SLS): Evidence for Safe Use in Household Cleaning Products," *Environ Health Insights* 9 (2015): 27–32.
4. D. Wharton, "Are Synthetic Playing Surfaces Hazardous to Athletes' Health? The Debate over 'Crumb Rubber' and Cancer," *Los Angeles Times,* February 28, 2016.
5. E. Menichini et al., "Artificial-Turf Playing Fields: Contents of Metals, PAH's, PCB's, PCDD's, and PCDF's, Inhalation Exposure to PAH's and Related Preliminary Risk Assessment," *Science of the Total Environment* 409, no. 23 (November 2011): 4950–57.
6. "The Total Audience Report: Q1 2016," Nielsen, June 27, 2016.
7. M. Wyde et al., "Report of Partial Findings from the National Toxicology Program Carcinogenesis Studies of Cell Phone Radiofrequency Radiation in HSD: Sprague Dawley® SD Rats (Whole Body Exposures)," *bioRxiv,* June 23, 2016.
8. J. D. Boice and R. E. Tarone, "Cell Phones, Cancer, and Children," *Journal of the National Cancer Institute* 103, no. 16 (August 2011): 1211–13; V. G. Khurana et al., "Cell Phones and Brain Tumors: A Review Including the Long-Term Epidemiologic Data," *Surgical Neurology* 72, no. 3 (September 2009): 205–14; S. Lagorio and M. Roosli, "Mobile Phone Use and Risk of Intracranial Tumors: A Consistency Analysis," *Bioelectromagnetics* 35, no. 2 (February 2014): 79–90; and S. Szmigielski, "Cancer Risks Related to Low-Level RF/MW Exposures, Including Cell Phones," *Electromagnetic Biology and Medicine* 32, no. 3 (September 2013): 273–80.
9. R. Feltman, "Do Cellphones Cause Cancer? Don't Believe the Hype," *Washington Post,* May 27, 2016.
10. J. D. Meeker, S. Sathyanarayana, and S. H. Swan, "Phthalates and Other Additives in Plastics: Human Exposure and Associated Health Outcomes," *Philosophical Transactions of the Royal Society B* 364, no. 1526 (July 2009): 2097–2113.
11. J. Glausiusz, "Toxicology: The Plastics Puzzle," *Nature* 508 (April 2014): 306–8.
12. L. Konkel, "Thermal Reaction: The Spread of Bisphenol S via Paper Products," *Environmental Health Perspectives* 121, no. 3 (March 2013).
13. R. U. Halden, "Plastics and Health Risks," *Annual Review of Public Health* 31 (April 2010): 179–94.

14. Jessica Chia, "The Truth About 'Fragrance-Free' Products," *Prevention,* January 23, 2014.

15. Halden, "Plastics and Health Risks," 179–94.

16. F. R. deGruijil, "Photocarcinogenesis: UVA vs. UVB," *Methods in Enzymology* 319, no. 2000 (December 2003): 359–66.

17. D. Gozal, "Sleep-Disordered Breathing and School Performance in Children," *Pediatrics* 102, no. 3 (September 1998): 616–20.

18. L. M. O'Brien et al., "Neurobehavioral Correlates of Sleep-Disordered Breathing in Children," *Journal of Sleep Research* 13, no. 2 (June 2004): 165–72.

19. K. M. Madsen et al., "A Population-Based Study of Measles, Mumps, and Rubella," *New England Journal of Medicine* 347 (November 2002): 1477–82.

20. "Blueberry Concentrate Improves Brain Function in Older People," *MDLinx,* March 8, 2017.

4. Get Me off Your F*cking Mailing List: A Study Worthy of Your Attention

1. David Mazières and Eddie Kohler, "Get Me Off Your Fucking Mailing List," PDF, Stanford Secure Computer Systems Group, Stanford University, retrieved November 22, 2014.

2. John Bohannon, "Who's Afraid of Peer Review?," *Science* 342, no. 6154 (October 2013): 60–65.

3. Gina Kolata, "A Scholarly Sting Operation Shines a Light on 'Predatory Journals,'" *New York Times,* March 22, 2017.

4. "Beall's List," Beall's List of Predatory Journals, December 2016.

5. Megan Molteni, "The FTC is Cracking Down on Predatory Science Journals," *Wired,* September 19, 2016.

6. A. J. Wakefield et al., "RETRACTED: Ileal-Lymphoid-Nodular Hyperplasia, Non-specific Colitis, and Pervasive Developmental Disorder in Children," *Lancet* 351, no. 9103 (February 1998): 637–41.

7. P. Sumner et al., "The Association Between Exaggeration in Health-Related Science News and Academic Press Releases: Retrospective Observational Study," *BMJ* 349 (December 2014): 1–8.

8. M. S. Pearce et al., "Radiation Exposure from CT Scans in Childhood and Subsequent Risk of Leukaemia and Brain Tumours: A Retrospective Cohort Study," *Lancet* 380, no. 9840 (August 2012): 499–505.

9. Jane Brody, "Ear Infection? Think Twice Before Inserting a Tube," *New York Times,* August 15, 2006.

10. https://en.wikipedia.org/wiki/Jahi_McMath_case.

11. Clay McNight, "Clinically Proven Weight-Loss Supplements," *Livestrong.com,* July 18, 2007.

12. T. Nagao, T. Itase, and I. Tokimitsu, "A Green Tea Extract High in Catechins Reduces Body Fat and Cardiovascular Risks in Humans," *Obesity* 15, no. 6 (June 2007): 1473–83.

13. Tom Gross, "'Clinically Proven' Doesn't Mean Much," *marinij.com*, June 16, 2008.

5. Tipping the Scale on a Balanced Diet:
You Are Not Always What You Eat

1. John LaRosa, "$65 Billion U.S. Weight Loss Market Is in Transition," *WebWire*, April 21, 2015.
2. D. L. Katz and S. Meller, "Can We Say What Diet Is Best for Health?," *Annual Review of Public Health* 35 (2014): 83–103.
3. "Statistics About Diabetes: Overall Numbers, Diabetes, and Prediabetes," www .diabetes.org, July 19, 2017.
4. Seth Schonwald, "Licorice Poisoning," *Medscape*, February 9, 2017.
5. University of Helsinki, "Pregnant Women Should Avoid Licorice," *Science News*, February 3, 2017.
6. G. Belakovic, "Mortality in Randomized Trials of Antioxidant Supplements for Primary and Secondary Prevention: Systematic Review and Meta-analysis," *JAMA* 297, no. 8 (February 2007): 842–57.
7. A. V. Klein and H. Kiat, "Detox Diets for Toxin Elimination and Weight Management: A Critical Review of the Evidence," *Journal of Human Nutrition Dietetics* 28, no. 6 (December 2015): 675–86.
8. https://en.wikipedia.org/wiki/Gluten.
9. Jacqueline Howard, "Gluten-Free Diets: Where Do We Stand?," *CNN.com*, March 10, 2017.
10. American Heart Association, "Low Gluten Diets Link to Higher Risk of Type 2 Diabetes," *Science News*, March 9, 2017.
11. https://en.wikipedia.org/wiki/FODMAP.
12. E. Lionetti et al., "Introduction of Gluten, HLA Status, and the Risk of Celiac Disease in Children," *New England Journal of Medicine* 371 (October 2014): 1295–1303.
13. Michael Specter, "Against the Grain: Should You Go Gluten-Free?," *New Yorker*, November 3, 2014.
14. R. H. M. Kwok, "Chinese Restaurant Syndrome," *New England Journal of Medicine* 278 (April 4, 1968): 796.
15. "The Simple Chemistry Behind Removing Wine Sulfites," *Wineoscope*, July 13, 2015.
16. S. Bunyavanich et al., "Peanut, Milk, and Wheat Intake During Pregnancy Is Associated with Reduced Allergy and Asthma in Children," *Journal of Allergy and Clinical Immunology* 133, no. 5 (May 2014): 1373–82.
17. G. D. Toit et al., "Randomized Trial of Peanut Consumption in Infants at Risk for Peanut Allergy," *New England Journal of Medicine* 372 (February 16, 2015): 803–13.
18. https://en.wikipedia.org/wiki/Genetically_modified_organism; A. Nicolia et al., "An Overview of the Last 10 Years of Genetically Engineered Crop Safety Research," *Critical Reviews in Biotechnology* 34, no. 1 (September 2013): 77–88; and

National Academies of Sciences, Engineering, and Medicine, "Agronomic and Environmental Effects of Genetically Engineered Crops," in *Genetically Engineered Crops: Experiences and Prospects,* 2016, 97–170.

19. http://www.who.int/foodsafety/areas_work/food-technology/faq-genetically-modified-food/en/.

20. M. Nestle, "Corporate Funding of Food and Nutrition Research: Science or Marketing?," *JAMA Internal Medicine* 176, no. 1 (January 2016): 6667.

21. C. Snell et al., "Assessment of the Health Impact of GM Plant Diets in Long-Term and Multigenerational Animal Feeding Trials: A Literature Review," *Food and Chemical Toxicology* 50, no. 3–4 (March–April 2012): 1134–48; and Nirvana Abou-Gabal, "Understanding the Controversy and Science of GMOs," *HuffingtonPost.com,* July 28, 2015.

22. Laura Ferguson, "The Bottom Line on Genetically Engineered Foods," *Tufts-Now,* May 24, 2016.

6. Fat-Free Sugar, Organic Cookies, and "Fresh" Produce: A Walk Through the Supermarket

1. C. M. Bulka et al., "The Unintended Consequences of a Gluten-Free Diet," *Epidemiology* 28, no. 3 (May 2017): e24–e25.

2. Allison Aubrey and Dan Charles, "Why Organic Food May Not Be Healthier for You," *Morning Edition,* NPR, September 4, 2012.

3. C. Smith-Spangler et al., "Are Organic Foods Safer or Healthier Than Conventional Alternatives? A Systematic Review," *Annals of Internal Medicine* 157, no. 5 (September 2012): 348–66.

4. Reynard Loki, "18 Fruits and Vegetables You Don't Need to Be Organic," *Salon,* June 23, 2015.

5. Alex Renton, "Why You Should *Never* Trust Labels on Food: 'Fresh' Food That Isn't Fresh, 'Natural' Colours That Are Chemicals, and 'Real' Fruit Juice That Is Only 5 Percent Fruit," *DailyMail.com,* September 1, 2010.

6. "Fish Faceoff: Wild Salmon vs. Farmed Salmon," Cleveland Clinic, https://health.clevelandclinic.org/2014/03/fish-faceoff-wild-salmon-vs-farmed-salmon/.

7. Tamar Hasnel, "Is Organic Better for Your Health? A Look at Milk, Meat, Eggs, Produce, and Fish," *Washington Post,* April 7, 2014.

8. Jennifer Welsh, "Genetically Engineered Salmon Is Perfectly Safe, FDA Says," *Business Insider,* December 28, 2012.

9. Alicia Mundy and Bill Tomson, "Eggs' 'Grade A' Stamp Isn't What It Seems," *Wall Street Journal,* September 2, 2010.

10. Anders Kelto, "Farm Fresh? Natural? Eggs Not Always What They're Cracked Up to Be," npr.org, December 23, 2014.

11. Tom Philpott, "How Factory Farms Play Chicken with Antibiotics," *Mother Jones,* May/June 2016.

12. J. Forman and J. Silverstein, "Organic Foods: Health and Environmental Advantages and Disadvantages," *Pediatrics* 130, no. 5 (November 2012): e1406–e1415.

13. J. Bradbury, "Docosahexaenoic Acid (DHA): An Ancient Nutrient for the Modern Brain," *Nutrients* 3, no. 5 (May 2011): 529–54.

14. P. Bozzatello et al., "Supplementation with Omega-3 Fatty Acids in Psychiatric Disorders: A Review of the Literature Data," *Journal of Clinical Medicine* 5, no. 8 (July 2016): 67.

15. Ibid.

16. Michelle Marinis, "Why Foods with Added DHA and ARA Should Scare You," *Mommynearest.com,* October 16, 2014.

17. Alexis Baden-Mayer, "GMO Ingredients in Organic Baby Food?," *Organicconsumers.org,* November 30, 2011.

18. P. Guesnet and J. M. Alessandri, "Docosahexaenoic Acid (DHA) and the Developing Central Nervous System (CNS)—Implications for Dietary Recommendations," *Biochimie* 93, no. 1 (January 2011): 772.

19. Alice Callahan, "Do DHA Supplements Make Babies Smarter?," *New York Times,* March 30, 2017.

20. Caroline Helwick, "Organic Foods Offer No Meaningful Nutritional Benefits, AAP Says," *Medscape,* October 24, 2012; Kristin Kiesel and Sofia B. Villas-Bous, "Got Organic Milk? Consumer Valuations of Milk Labels After the Implementation of the USDA Organic Seal," *DeGruyter.com,* April 16, 2007; and Forman and Silverstein, "Organic Foods," e1406–e1415.

21. http://www.fda.gov/Food/GuidanceRegulation/GuidanceDocumentsRegulatory Information/Labeling/Nutrition/ucm064916.htm.

22. Jen Gunter, "Will I Get a Yeast Infection If I Eat Too Much Sugar?," posted on her blog site, December 26, 2011, https://drjengunter.wordpress.com/2011/12/26/will-i-get-a-yeast-infection-if-i-eat-too-much-sugar/.

23. Carl Lavie, *The Obesity Paradox: When Thinner Means Sicker and Heavier Means Healthier* (New York: Hudson Street Press, 2014).

7. The True Cost of Being Fortified: Supplements, Powders, and Potions

1. E. Guallar et al., "Enough Is Enough: Stop Wasting Money on Vitamin and Mineral Supplements," *Annals of Internal Medicine* 159, no. 2 (December 2013): 850–51. For more on this study go to www.hopkinsmedicine.org.

2. https://www.cancer.gov/about-cancer/causes-prevention/risk/diet/antioxidants-fact-sheet.

3. G. Drouin, J. R. Godin, and B. Page, "The Genetics of Vitamin C Loss in Vertebrates," *Current Genomics* 12, no. 5 (August 2011): 371–78.

4. Carl Zimmer, "Learning from the History of Vitamins," *New York Times,* December 12, 2013.

5. Carl Zimmer, "Vitamins' Old, Old Edge," *New York Times,* December 9, 2013.

6. Paul Offit, "The Vitamin Myth: Why We Think We Need Supplements," *Atlantic,* July 19, 2013.

7. Ibid.

8. *Linus Pauling in His Own Words: Selections from His Writings, Speeches, and Interviews*, ed. Barbara Marinacci (New York: Simon & Schuster, 1995).

9. G. Lippi and M. Franchini, "Vitamin K in Neonates: Facts and Myths," *Blood Transfusion* 9, no. 1 (January 2011): 4–9.

10. Policy Statement, Committee on Fetus and Newborn, "Controversies Concerning Vitamin K and the Newborn," *Pediatrics* 112, no. 1 (July 2003): 191–92; and R. Schulte et al., "Rise in Late Onset Vitamin K Deficiency Bleeding in Young Infants Because of Omission or Refusal of Prophylaxis at Birth," *Pediatric Neurology* 50, no. 6 (June 2014): 564–68.

11. https://www.cdc.gov/ncbddd/vitamink/olive-story.html.

12. B. M. P. Tang et al., "Use of Calcium or Calcium in Combination with Vitamin D Supplementation to Prevent Fractures and Bone Loss in People Aged 50 Years and Older: A Meta-analysis," *Lancet* 370, no. 9588 (August 2007): 657–66.

13. https://uspreventiveservicestaskforce.org/Page/Document/Recommendation StatementFinal/vitamin-d-and-calcium-to-prevent-fractures-preventive -medication#consider.

14. J. Wactawski-Weode et al., "Calcium plus Vitamin D Supplementation and the Risk of Colorectal Cancer," *New England Journal of Medicine* 354 (February 2006): 684–96.

15. Richard Knox, "How a Vitamin D Test Misdiagnosed African Americans," npr .org, November 20, 2013.

16. https://www.cdc.gov/ncbddd/folicacid/recommendations.html; and https://www .acog.org/Patients/FAQs/Nutrition-During-Pregnancy#much.

17. http://www.fda.gov/Drugs/DevelopmentApprovalProcess/HowDrugsare DevelopedandApproved/default.htm.

18. "Should You Take Dietary Supplements?: A Look at Vitamins, Minerals, Botanicals, and More," *NIH News in Health,* August 2013.

8. Raise Your Glass: Water, Water, Everywhere

1. R. Wolf et al., "Nutrition and Water: Drinking Eight Glasses of Water a Day Ensures Proper Skin Hydration—Myth or Reality?," *Clinical Dermatology* 28, no. 4 (July–August 2010): 380–3.

2. Rachel C. Vreeman and Aaron E. Carroll, "Medical Myths," *BMJ* 335, no. 7633 (2007): 1288–89.

3. Tara Parker-Pope, "Medical Myths Even Doctors Believe," *New York Times,* December 26, 2007, https://well.blogs.nytimes.com/2007/12/26/medical-myths -even-doctors-believe/?hp&apage=2.

4. Johanna R. Rochester and Ashley L. Bolden, "Bisphenol S and F: A Systematic Review and Comparison of the Hormonal Activity of Bisphenol A Substitutes," *Environmental Health Perspectives* 123, no. 7 (July 2015): 643–50; and Jenna Bilbrey, "BPA-Free Plastic Containers May Be Just as Hazardous," *Scientific*

American, August 11, 2014, https://www.scientificamerican.com/article/bpa-free-plastic-containers-may-be-just-as-hazardous/.

5. R. Rezg et al., "Bisphenol A and Human Chronic Diseases: Current Evidences, Possible Mechanisms, and Future Perspectives," *Environment International* 64 (March 2014): 83–90.

6. M. Eriksen et al., "Plastic Pollution in the World's Oceans: More Than 5 Trillion Plastic Pieces Weighing over 250,000 Tons Afloat at Sea," *PLoS ONE* 9, no. 12 (2014): e111913.

7. https://ofmpub.epa.gov/apex/safewater/f?p=136:102; and Sanaz Majd, "Should You Drink Tap or Bottled Water?," *Scientific American,* October 21, 2015, https://www.scientificamerican.com/article/should-you-drink-tap-or-bottled-water/.

8. Patrick Allan, "Three Myths About Sparkling Water, Debunked," *Lifehacker.com,* February 16, 2016, http://lifehacker.com/three-myths-about-sparkling-water-debunked-1759280798.

9. E. Gonzalez de Mejia and M. V. Ramirez-Mares, "Impact of Caffeine and Coffee on Our Health," *Trends in Endocrinology & Metabolism* 25, no. 10 (October 2014): 489–92.

10. D. C. Mitchell et al., "Beverage Caffeine Intakes in the U.S.," *Food and Chemical Toxicology* 63 (January 2014): 136–42.

11. "The Buzz on Energy-Drink Caffeine," *Consumer Reports,* December 2012, http://www.consumerreports.org/cro/magazine/2012/12/the-buzz-on-energy-drink-caffeine/index.htm.

12. Rachel Bachman, "Caffeine: The Performance Enhancer in Your Kitchen," *Wall Street Journal,* July 25, 2016, http://www.wsj.com/articles/caffeine-the-performance-enhancer-in-your-kitchen-1469457168.

13. Christina J. Calamaro, Thornton B. A. Mason, and Sarah J. Ratcliffe, "Adolescents Living the 24/7 Lifestyle: Effects of Caffeine and Technology on Sleep Duration and Daytime Functioning," *Pediatrics* 123, no. 6 (June 2009).

14. Veronica Hackethal, "Liver Cancer Report: Obesity and Alcohol Up Risk," News & Perspectives, *Medscape,* March 26, 2015, http://www.medscape.com/viewarticle/842122.

15. https://en.wikipedia.org/wiki/Nurses'_Health_Study.

16. http://www.nurseshealthstudy.org/sites/default/files/pdfs/table%20v2.pdf.

17. Patrick J. Skerrett, "Resveratrol—the Hype Continues," *Harvard Health Blog,* Harvard Health Publications, February 3, 2012, http://www.health.harvard.edu/blog/resveratrol-the-hype-continues-201202034189.

18. S. J. Park et al., "Resveratrol Ameliorates Aging-Related Metabolic Phenotypes by Inhibiting cAMP Phosphodiesterases," *Cell* 148, no. 3 (February 2012): 421–33.

19. J. H. O'Keefe et al., "Alcohol and Cardiovascular Health: The Dose Makes the Poison . . . or the Remedy," *Mayo Clinic Proceedings* 89, no. 3 (March 2014): 382–93.

9. Putting the *C* Back in CAM: Complementary Alternative Medicine

1. https://www.cancer.gov/about-cancer/treatment/cam; and https://nccih.nih.gov/health/cancer/complementary-integrative-research.
2. http://www.mayoclinic.org/healthy-lifestyle/consumer-health/in-depth/alternative-medicine/art-20045267.
3. http://www.newlifemedicalclinics.com/.
4. http://www.anoasisofhealing.com/.
5. http://www.burzynskiclinic.com/.
6. Peter Lipson, "FDA Documents Paint Disturbing Picture of Burzynski Cancer Clinic," *Forbes,* November 11, 2013, https://www.forbes.com/sites/peterlipson/2013/11/11/fda-documents-paint-disturbing-picture-of-burzynski-cancer-clinic/#633590a26087.
7. http://atavisticchemotherapy.com/.
8. http://atavisticchemotherapy.com/atavistic-chemotherapy/atavistic-chemotherapy-proof-of-concept-and-clinical-validation/.
9. Eula Biss, *On Immunity: An Inoculation* (Minneapolis: Graywolf Press, 2014).
10. Much of the section "Mesmerized by Homeopathy" was synthesized from historical information featured in S. H. Podolsky and A. S. Kesselheim, "Regulating Homeopathic Products—a Century of Dilute Interest," *New England Journal of Medicine* 374, no. 3 (January 2016): 201–3.
11. Ibid.
12. https://nccih.nih.gov/research/statistics/NHIS/2012/key-findings.
13. FDA, "Warnings on Three Zicam Intranasal Zinc Products," Consumer Updates, June 16, 2009, https://www.fda.gov/ForConsumers/ConsumerUpdates/ucm166931.htm.
14. Oliver Wendell Holmes, *Homeopathy and Its Kindred Delusions: Two Lectures Delivered Before the Boston Society for the Diffusion of Useful Knowledge* (Boston: William D. Ticknor, 1842).
15. Richard Dawkins, *A Devil's Chaplain: Reflections on Hope, Lies, Science, and Love* (New York: Houghton Mifflin Harcourt, 2003).
16. Federal Trade Commission, "FTC Issues Enforcement Policy Statement Regarding Marketing Claims for Over-the-Counter Homeopathic Drugs," press release, November 15, 2016, https://www.ftc.gov/news-events/press-releases/2016/11/ftc-issues-enforcement-policy-statement-regarding-marketing.
17. Paul Glasziou, "Still No Evidence for Homeopathy," *BMJ,* February 16, 2016, http://blogs.bmj.com/bmj/2016/02/16/paul-glasziou-still-no-evidence-for-homeopathy/.
18. Maj-Britt Niemi, "Placebo Effect: A Cure in the Mind," *Scientific American,* February 2009, https://www.scientificamerican.com/article/placebo-effect-a-cure-in-the-mind/.
19. Mallika Marshall, "A Placebo Can Work Even When You Know It's a Placebo,"

Harvard Health Blog, Harvard Health Publications, July 7, 2016, http://www
.health.harvard.edu/blog/placebo-can-work-even-know-placebo-201607
079926.

20. Robin Holtedahl, Jens Ivar Brox, and Ole Tjomsland, "Placebo Effects in Trials
Evaluating 12 Selected Minimally Invasive Interventions: A Systematic Review
and Meta-analysis," *BMJ Open* 5, no. 1 (2015): e007331.

21. T. J. Kaptchuk and F. G. Miller, "Placebo Effects in Medicine," *New England
Journal of Medicine* 373, no. 1 (July 2015): 8–9.

10. Take a Shot: The Perils of Losing the Herd

1. Parts of this section were derived from an article I wrote for *The Hollywood
Reporter*'s Guest Column: Nina Shapiro, "Measles Hit Hollywood amid Vaccination Battle: Doctor Addresses 'Grave and Sad Situation,'" *Hollywood
Reporter,* February 19, 2015, http://www.hollywoodreporter.com/news/mea
sles-hit-hollywood-vaccination-battle-775270?.

2. https://www.cdc.gov/flu/asthma/.

3. Josef Neu and Jona Rushing, "Cesarean Versus Vaginal Delivery: Long-Term Infant Outcomes and the Hygiene Hypothesis," *Clinics in Perinatology* 38, no. 2
(2011): 321–31.

4. https://www.biomedcentral.com/about/press-centre/science-press-releases/17
-nov-2014.

5. S. E. Gould, "The Bacteria in Breast Milk," *Scientific American,* December 8,
2013, https://blogs.scientificamerican.com/lab-rat/the-bacteria-in-breast-milk/.

6. X. Huang et al., "Mercury Poisoning: A Case of a Complex Neuropsychiatric
Illness," *American Journal of Psychiatry* 171, no. 12 (December 2014): 1253–56.

7. J. G. Dorea, "Mercury and Lead During Breast-Feeding," *British Journal of
Nutrition* 92, no. 1 (July 2004): 21–40.

8. Helen Petousis-Harris, "Myths Surrounding Vaccines," in *The Practical Compen-
dium of Immunisations for International Travel,* ed. Marc Shaw and Claire Wong
(New York: Adis, 2015), 175–79.

9. Ana Clara Monsalvo et al., "Severe Pandemic 2009 H1N1 Influenza Disease due
to Pathogenic Immune Complexes," *Nature Medicine,* 2010. See also Vanderbilt
University Medical Center, "Over-Reactive Immune System Kills Young Adults
During Pandemic Flu," *ScienceDaily,* accessed September 1, 2017, www.science
daily.com/releases/2010/12/101205202526.htm.

10. Ari Brown, "Clear Answers and Smart Advice About Your Baby's Shots," Immunization Action Coalition, Saint Paul, Minn., http://www.immunize.org
/catg.d/p2068.pdf.

11. http://vaccine-safety-training.org/live-attenuated-vaccines.html.

12. https://www.cdc.gov/tetanus/about/symptoms-complications.html.

13. https://www.usnews.com/dbimages/master/8226/GR_PR_081203Vaccines
.png',870,400.

14. Paul A. Offit and Charlotte A. Moser, "The Problem with Dr. Bob's Alternative Vaccine Schedule," *Pediatrics* 123, no. 1 (January 2009).

15. Bob Sears, "The Truth About Vaccines and Autism," *iVillage*, September 2009.

16. F. DeStefano, C. S. Price, and E. S. Weintraub, "Increasing Exposure to Antibody-Stimulating Proteins and Polysaccharides in Vaccines Is Not Associated with Risk of Autism," *Journal of Pediatrics* 163, no. 2 (August 2013): 561–67.

17. Bourree Lam, "Vaccines Are Profitable, So What?," *Atlantic,* February 10, 2015, https://www.theatlantic.com/business/archive/2015/02/vaccines-are-profitable-so -what/385214/.

18. https://www.cdc.gov/vaccinesafety/research/iomreports/index.html.

19. Brown, "Clear Answers and Smart Advice."

20. https://www.cancer.gov/about-cancer/understanding/statistics.

21. American Academy of Pediatrics, "HPV Vaccination Does Not Lead to Increased Sexual Activity," press release, October 15, 2012, https://www.aap.org/en-us /about-the-aap/aap-press-room/Pages/HPV-Vaccination-Does-Not-Lead-to -Increased-Sexual-Activity.aspx.

11. Testing, Testing, One, Two, Three: From the Outside Looking In

1. The "thirty-five thousand" decisions a day calculation is widely cited, though it is difficult to find an original source. See J. Sollisch, "The Cure for Decision Fatigue," *Wall Street Journal,* June 11–12, 2016.

2. Ananya Mandal, "Huntington's Disease History," *News-Medical.net,* last updated September 11, 2014, http://www.news-medical.net/health/Huntingtons -Disease-History.aspx.

3. Aaron E. Carroll and Austin Frakt, "How to Measure a Medical Treatment's Potential for Harm," *New York Times,* February 2, 2015, https://www.nytimes .com/2015/02/03/upshot/how-to-measure-a-medical-treatments-potential-for -harm.html?_r=0.

4. P. M. Rothwell et al., "Effect of Daily Aspirin on Long-Term Risk of Death due to Cancer: Analysis of Individual Patient Data from Randomised Trials," *Lancet* 377, no. 9759 (January 2011): 31–41.

5. J. Cuzick et al., "Estimates of Benefits and Harms of Prophylactic Use of Aspirin in the General Population," *Annals of Oncology* 26, no. 1 (January 2015): 47–57.

6. https://www.cancer.org/healthy/find-cancer-early/cancer-screening-guidelines /american-cancer-society-guidelines-for-the-early-detection-of-cancer.html.

7. J. G. Elmore et al., "Diagnostic Concordance Among Pathologists Interpreting Breast Biopsy Specimens," *JAMA* 313, no. 11 (March 2015): 1122–32.

8. Tom Balfour, "Cope's Early Diagnosis of the Acute Abdomen," *Journal of the Royal Society of Medicine* 99, no. 1 (2006): 42.

9. C. A. Coursey et al., "Making the Diagnosis of Acute Appendicitis: Do More Preoperative CT Scans Mean Fewer Negative Appendectomies? A 10-Year Study," *Radiology* 254, no. 2 (February 2010): 460–68.

10. A. S. Raja et al., "Negative Appendectomy Rate in the Era of CT: An 18-Year Perspective," *Radiology* 256, no. 2 (August 2010): 460–65.
11. K. K. Varadhan, K. R. Neal, and D. N. Lobo, "Safety and Efficacy of Antibiotics Compared with Appendicectomy for Treatment of Uncomplicated Acute Appendicitis: Meta-analysis of Randomised Controlled Trials," *BMJ* 344 (April 2012): e2156.
12. U.S. Preventive Services Task Force, "Screening for Colorectal Cancer U.S. Preventive Services Task Force Recommendation Statement," *JAMA* 315, no. 23 (2016): 2564–75.
13. A. Chukmaitov, et al., "Patient Comorbidity and Serious Adverse Events after Outpatient Colonoscopy: Population-based Study From Three States, 2006 to 2009," *Dis Colon Rectum* 59, no. 7 (July 2016): 677–87.
14. J. M. Inadomi, "Colorectal Cancer Screening: Which Test Is Best?," *JAMA Oncology* 2, no. 8 (2016): 1001–3.
15. A. Jemal et al., "Prostate Cancer Incidence and PSA Testing Patterns in Relation to USPSTF Screening Recommendations," *JAMA* 314, no. 19 (2015): 2054–61.

12. When 50 Is the New 40: Drinking from the Fountain of Youth

1. Erika Check Hayden, "Anti-Ageing Pill Pushed as Bona Fide Drug," *Nature,* June 17, 2015, http://www.nature.com/news/anti-ageing-pill-pushed-as-bona -fide-drug-1.17769.
2. Andrew Sussman, "What Happens When a Retail Pharmacy Decides to Stop Selling Cigarettes?," *HealthAffairs Blog,* February 26, 2015, http://healthaffairs .org/blog/2015/02/26/what-happens-when-a-retail-pharmacy-decides-to-stop -selling-cigarettes/.
3. Ruta Ganceviciene et al., "Skin Anti-Aging Strategies," *Dermato-Endocrinology* 4, no. 3 (2012): 308–19.
4. http://www.mayoclinic.org/diseases-conditions/wrinkles/in-depth/wrinkle -creams/art-20047463?pg=2.
5. http://www.skincancer.org/skin-cancer-information/skin-cancer-facts.
6. Steven Q. Wang, "Does a Higher SPF (Sun Protection Factor) Sunscreen Always Protect Your Skin Better?," Ask the Expert, Skin Cancer Foundation, http:// www.skincancer.org/skin-cancer-information/ask-the-experts/does-a-higher -spf-sunscreen-always-protect-your-skin-better.
7. https://www.cdc.gov/botulism/testing-treatment.html.
8. P. K. Nigam and Anjana Nigam, "Botulinum Toxin," *Indian Journal of Dermatology* 55, no. 1 (January–March 2010): 8–14.
9. P. J. Snyder et al., "Effects of Testosterone Treatment in Older Men," *New England Journal of Medicine* 374, no. 7 (February 2016): 611–24.
10. P. J. Jenkins, A. Mukherjee, and S. M. Shalet, "Does Growth Hormone Cause Cancer?," *Clinical Endocrinology* (Oxf) 64, no. 2 (February 2006): 115–21.
11. Danny Eapen et al., "Raising HDL Cholesterol in Women," *International Journal of Women's Health* 1 (2009): 181–91; and S. Lamon-Fava et al., "Role of the

Estrogen and Progestin in Hormonal Replacement Therapy on Apolipoprotein A-I Kinetics in Postmenopausal Women," *Arteriosclerosis, Thrombosis, and Vascular Biology* 26, no. 2 (February 2006): 385–91.

12. https://www.cancer.gov/about-cancer/causes-prevention/risk/hormones/mht-fact-sheet.

13. https://www.nih.gov/health-information/menopausal-hormone-therapy-information.

14. D. R. Pachman, J. M. Jones, and C. L. Loprinzi, "Management of Menopause-Associated Vasomotor Symptoms: Current Treatment Options, Challenges and Future Directions," *International Journal of Women's Health* 2 (August 2010): 123–35.

15. https://www.nhlbi.nih.gov/whi/index.html.

16. https://www.nhlbi.nih.gov/whi/whi_faq.htm.

17. H. W. Chae, D. H. Kim, and H. S. Kim, "Growth Hormone Treatment and Risk of Malignancy," *Korean Journal of Pediatrics* 58, no. 2 (February 2015): 41–46.

18. Andrzej Bartke, "Growth Hormone and Aging: A Challenging Controversy," *Clinical Interventions in Aging* 3, no. 4 (December 2008): 659–65.

19. D. Rudman et al., "Effects of Human Growth Hormone in Men over 60 Years Old," *New England Journal of Medicine* 323, no. 1 (July 1990): 1–6.

20. Bartke, "Growth Hormone and Aging," 659–65.

21. http://www.webmd.com/fitness-exercise/human-growth-hormone-hgh#2.

22. A. Sifferlin, "How Silicon Valley Is Trying to Hack Its Way to a Much (Much, Much) Longer Life," *Time,* February 27–March 6, 2017.

23. https://stemcells.nih.gov/info/basics/4.htm.

24. Zoe Corbyn, "Live Forever: Scientists Say They'll Soon Extend Life 'Well Beyond 120,'" *Guardian,* January 11, 2015, https://www.theguardian.com/science/2015/jan/11/-sp-live-forever-extend-life-calico-google-longevity.

25. https://www.ncbi.nlm.nih.gov/pubmed/26512657.

13. Hyped Exercise: Climb Every Mountain

1. C. S. Almond et al., "Hyponatremia Among Runners in the Boston Marathon," *New England Journal of Medicine* 352, no. 15 (April 2005): 1550–56.

2. D. R. Bassett, P. L. Schneider, and G. E. Huntington, "Physical Activity in an Old Order Amish Community," *Medicine & Science in Sports & Exercise* 36, no. 1 (January 2004): 79–85.

3. C. Tudor-Locke and D. R. Bassett Jr., "How Many Steps/Day Are Enough? Preliminary Pedometer Indices for Public Health," *Sports Medicine* 34, no. 1 (2004): 1–8.

4. Jesse Singal, "How Many Steps a Day Should You Really Walk?," *New York,* June 5, 2015, http://nymag.com/scienceofus/2015/06/how-many-steps-a-day-really-walk.html#jumpLink.

5. Aaron E. Carroll, "Wearable Fitness Devices Don't Seem to Make You Fitter," *New York Times,* February 20, 2017; and http://jamanetwork.com/journals/jama/fullarticle/2553448.

6. Robinson Meyer, "The Quantified Welp," *Atlantic,* February 25, 2016, https://www.theatlantic.com/technology/archive/2016/02/the-quantified-welp/470874/?utm_source=atlfb.

7. F. El-Amrawy and M. I. Nounou, "Are Currently Available Wearable Devices for Activity Tracking and Heart Rate Monitoring Accurate, Precise, and Medically Beneficial?," *Healthcare Informatics Research* 21, no. 4 (October 2015): 315–20.

8. Tara Siegel Bernard, "Giving Out Private Data for Discount in Insurance," *New York Times,* April 8, 2015, https://www.nytimes.com/2015/04/08/your-money/giving-out-private-data-for-discount-in-insurance.html?_r=0.

9. Kristi King and Ann Swank, "Exercise Strategies for Children: A Public Health Approach for Obesity Prevention," *ACSM's Health and Fitness Journal* 19, no. 4 (2015): 39–41.

10. J. P. DiFiori et al., "Overuse Injuries and Burnout in Youth Sports: A Position Statement from the American Medical Society for Sports Medicine," *British Journal of Sports Medicine* 48 (2014): 287–88.

11. Alice Part, "Extreme Workouts: When Exercise Does More Harm Than Good," *Time,* June 4, 2012, http://healthland.time.com/2012/06/04/extreme-workouts-when-exercise-does-more-harm-than-good/.

12. M. Brogan et al., "Freebie Rhabdomyolysis: A Public Health Concern. Spin Class–Induced Rhabdomyolysis," *American Journal of Medicine* 130, no. 4 (April 2017): 484–87.

13. http://www.yogabasics.com/learn/history-of-yoga/.

14. Jordan Shakeshaft, "The 17 Biggest Fitness Fads That Flopped," Greatest.com, January 30, 2012, http://greatist.com/fitness/17-biggest-fitness-fads-flopped.

15. A. Biswas et al., "Sedentary Time and Its Association with Risk for Disease Incidence, Mortality, and Hospitalization in Adults: A Systematic Review and Meta-analysis," *Annals of Internal Medicine* 162 (2015): 123–32.

16. http://www.who.int/topics/physical_activity/en/.

17. Rachel Krantz, "Are Standing Desks Really Healthier? 8 Things You Should Know Before You Renounce Your Chair," Bustle.com, March 3, 2016, https://www.bustle.com/articles/144401-are-standing-desks-really-healthier-8-things-you-should-know-before-you-renounce-your-chair.

Conclusion: Don't Believe the Hype: Is It All Hype?

1. Sarah Maslin Nir and Nate Schweber, "After 146 Years, Ringling Brothers Circus Takes Its Final Bow," *New York Times,* May 21, 2017, https://www.nytimes.com/2017/05/21/nyregion/ringling-brothers-circus-takes-final-bow.html.

2. http://hoaxes.org/archive/permalink/joice_heth; and http://hoaxes.org/archive/permalink/the_feejee_mermaid.

3. https://www.youtube.com/watch?v=3qZA-xOeQmE; and Nir and Schweber, "After 146 Years."

4. Eric Topol, "The Smart-Medicine Solution to the Health-Care Crisis," *Wall Street Journal,* July 7, 2017, https://www.wsj.com/articles/the-smart-medicine-solution-to-the-health-care-crisis-1499443449.

5. Roni Caryn Rabin, "Is Alcohol Good for You? An Industry-Backed Study Seeks Answers," *New York Times,* July 3, 2017, https://www.nytimes.com/2017/07/03/well/eat/alcohol-national-institutes-of-health-clinical-trial.html?emc=edit_th_20170704&nl=todaysheadlines&nlid=22330961.

INDEX